SpringerBriefs in Energy

More information about this series at http://www.springer.com/series/8903

Rui F.M. Lobo

Nanophysics for Energy Efficiency

 Springer

Rui F.M. Lobo
Physics Department
Universidade Nova de Lisboa - Faculdade de
    Ciências e Tecnologia
Caparica
Portugal

ISSN 2191-5520   ISSN 2191-5539   (electronic)
SpringerBriefs in Energy
ISBN 978-3-319-17006-0   ISBN 978-3-319-17007-7   (eBook)
DOI 10.1007/978-3-319-17007-7

Library of Congress Control Number: 2015945132

Springer Cham Heidelberg New York Dordrecht London

Printed on acid-free paper

Springer International Publishing AG Switzerland is part of Springer Science+Business Media
(www.springer.com)

For their moral support, I dedicate this book to my close family members, particularly those who are the reason for my existence, and to my soul mate, and offspring in charge to ensure the temporal continuity of part of my DNA.

Foreword

Soon after the invention of the Scanning Probe Microscopies family (STM and AFM), an AFM was modified to be able to measure friction forces. Subsequent developments led to their use in studies of scratching, wear, and indentation and measurements of mechanical and electrical properties. Now they are routinely used in many fields including micro/nanotribology. Before becoming involved in Scanning Probe Microscopies (SPMs), the author had carried out projects in Nanophysics, in particular single-molecule experiments and weak interactions from quantum nature, which include molecular beams and nanoscopic/spectroscopic analyses. Collaboration of Nanotribology and Scanning Local Probes was timely expected and in 1996, I organized with the author the NATO Study Advanced Institute on Micro/Nanotribology in Sesimbra, Portugal, where important and interesting results were presented.

The development of this field has attracted numerous physicists and SPMs in Nanotribology have made an immense impact on the field of Advanced Nanotechnology. As an example, studies with SPMs involve properties and size scales of critical relevance to energy-related components.

As a pioneer of Nanotribology I must say that advanced local probe techniques and in particular emergent scanning probe microscopy methods addressing several nanoscale functionalities have been gradually implemented in energy storage and conversion applications. Friction is also an important limitation of energy efficiency performances in electro-mechanical components and some SPM techniques already are used for electric energy storage studies to study nano-friction.

More recently, local probe techniques associated with nanomachines are expected to provide relevant information about the fundamental nature of space and friction, especially concerning the zero-point electromagnetic field.

I wish to congratulate Professor Rui Lobo in publishing this very timely book entitled *Advanced Topics on Nanophysics for Energy Efficiency* which explores the important bridging between Nanophysics and Energy fields. I expect that it will be well received by the international scientific community.

USA   Bharat Bhushan
January 2015   Ohio Eminent Professor and
                                                       The Howard D. Winbigler Professor
                                                     Director, Nanoprobe Laboratory for
                                            Bio- & Nanotechnology and Biomimetics

Preface

Ordinary human experience spans a range of space scales from meters to thousands of kilometers, the longest lengths a mere billion or so times larger than the shortest. The progress of nanophysics and nanotechnology has been marked by the scientist's growing familiarity with sizes that are very much shorter than those that are experienced in our lives.

Around 465 BC, Democritus' atomic hypothesis considered that all matter consists of invisible solid particles called atoms, homogeneous, indestructible, and can differ in size and shape. Democritus had postulated the first serious estimation of a length much shorter than a human naked eye observation and later, in the Modern Era, the atomic postulates of chemistry were enough to explain quantitatively several transformations of matter. However, it was necessary to wait till the twentieth century for decisive experimental evidences of atomic and molecular realities and their interpretation by quantum physics. This century, due to the enormous progress in experimental physics, we became used to the lowermost length scales and thanks to the recent development of advanced scanning local probes we are now familiar with images of atoms and molecules as real entities. Even more surprisingly, we are now in possession of the tools that allow us to control with a high precision the manipulation of these entities at the nanoscale.

It is remarkable that our experiments and theories have become sufficiently reliable so that we can now get pictures at the nanometer scale with some confidence that we know what we are talking about.

Since the 2000 Nanotechnology Initiative announced in the US, the financial resources for Nanophysics and Nanotechnology have been increased all over the world, and even some years earlier I already had led work in Nanophysics, in particular single-particle experiments. In my group, projects in single-molecule experiments and weak interactions, which include molecular beams and nanoscopic/spectroscopic analysis have been conducted (electron transfer dynamics in atom-polyatomic molecule collisions and negative fullerene conformers worth a mention).

My privilege of having attended the First Advanced School on Scanning Tunneling Microscopy organized by the Nobel laureate in Physics H. Rohrer, enabled my group over the following years to be involved in the development and training of various techniques belonging to the SPM family (STM, AFM, SNOM).

Before this enthusiastic outlook, a collaboration in Nanotribology and Scanning Local Probes was timely expected and in 1996 Prof. Bharat Bhushan organized with myself the NATO Study Advanced Institute on Micro/Nanotribology (Sesimbra-Portugal), where important and interesting results covering many areas of this field were presented. Soon after the invention of the initial relevant members of the Scanning Probe Microscopies family (STM, AFM) it was discovered that part of the information in the images resulted from friction and so AFM in particular could be used as a tool for Nanotribology. With the advent of more powerful computers, atomic scale simulations have been able to predict the observed phenomena. Development of the field Micro/Nanotribology has attracted numerous physicists and SPMs in Nanotribology have an immense impact in the field of Advanced Nanotechnology for Energy and Ultra-sensitive Detection of several interactions. Tribology is the science and technology of two interacting surfaces in relative motion and Micro/Nanotribology studies are needed to develop a fundamental understanding of interfacial phenomena in Micro/Nano-electromechanical Systems (MEMS/NEMS).

On the other hand, the advances in scanning probe microscopy (SPM) involve properties and size scales of critical relevance to energy-related materials. The design and control of materials at the nanoscale are the foundation of many new strategies for energy generation, storage, and efficiency.

So this book starts to address scanning probe microscopy (SPM) methods in several nanoscale functionalities that have been successfully implemented in energy storage and conversion applications. Novel phenomena and properties have been explored and revealed. Such studies on some electric energy storage devices is bridging our gaps in the understanding of some local phenomena, such as aging at the nanoscale.

This approach is opening new aspects of Nanophysics and Nanotechnology which can never be attained by studies using non in-situ and non-local techniques. Mainly the author deals here with Scanning Probe Microscopies but the topics are not limited to them.

An important viewpoint here is to focus the reader's attention on the original concepts and ideas that can be achieved by studying the physics of Scanning Probe Microscopies applied to the study of electric energy storage devices.

Thankfully, all the results described here reflect a clear motivation toward sustainable energy applications, in particular, energy efficiency.

This research book integrates knowledge from both the energy science and instrumentation points of view and is intended for two types of readers: postgraduate students in Nanophysics and Nanotechnology and researchers in academia who intended to become active in the field of Nanophysics for Energy. It should also serve as a good text for postgraduate courses in Advanced Nanotechnology and Applied Nanophysics.

New University of Lisbon, 2015   Rui F.M. Lobo

Acknowledgments

- For his encouragement in bringing this book to life, I wish to thank Bharat Bhushan, Eminent Professor at the Ohio State University (USA).
- The author gratefully acknowledges FCT-MCTES—Portugal for the grant ref. SFRH/BSAB/1355/2013 and the project PEST (UID/EEA/00066/2013).

New University of Lisbon, 2015   Rui F.M. Lobo

Contents

Abbreviations

AFM	Atomic Force Microscopy
APMS	Atom Probe Mass Spectrometry
APTES	Aminopropyltriethoxysilane
cAFM	Conductive Atomic Force Microscopy
CNT	Carbon Nanotube
CP-AFM	Conducting-probe Atomic Force Microscopy
CPD	Contact Potential Difference
CSAFM	Current Sensing AFM
CV	Cyclic Voltammetry
DBCNPs	Double Carbon Nanoprobes
DFT	Density Functional Theory
DME	Dimethoxyethane
EC-STM	Electrochemical Scanning Tunneling Microscopy
EDL	Electrochemical Double Layer Capacitor
EDLC	Electric Double Layer Capacitor
EES	Electric Energy Storage
EFM	Electric Force Microscopy
EIS	Electrochemical Impedance Spectroscopy
ESM	Electrochemical Strain Microscopy
ESPM	Electrical Scanning Probe Microscopy
EWF	Electronic Work Function
FC	Fuel Cell
FFM	Friction Force Microscopy
F-SNOM	Fluorescence SNOM
FTIR	Fourier Transform Infrared Spectroscopy
GC	Galvanostatic Cycling
HOMO	Highest Occupied Molecular Orbital
HOPG	Highly Oriented Pyrolytic Graphite
HSAC	High Surface Area Carbon
HV	High-Voltage
IR	Infrared

ITO	Indium Tin Oxide
IV	Current-Voltage
KPM	Kelvin Probe Microscopy
LIBs	Lithium-ion batteries
LUMO	Lowest Unoccupied Molecular Orbital
MB-TDS	Molecular Beam Thermal Desorption Spectrometry
MEMS	Micro-electromechanical Systems
MFM	Magnetic Force Microscopy
MWNT	Multi-walled Carbon Nanotubes
NC-AFM	Non-contact AFM
NCD	Nano-crystalline Deposit
NEMS	Nano-electromechanical Systems
NFDM	Near-field Dielectric Microscopy
NFIS	Near-field Infrared Spectrometry
NFS	Near-Field Spectroscopy
ORR	Oxygen Reduction Reaction
OTS	Octadecyltrichlorosilane
pcAFM	Photoconductive Atomic Force Microscopy
PET	Polyethylene Terephthalate
PVDF	Polyvinylidene Difluoride
PZT	Piezo Tube
RMS	Root Mean Square
RS	Raman Spectroscopy
R-SNOM	SNOM Raman
SAMs	Self-assembled Monolayers
SECM	Scanning Electrochemical Microscopy
SEI	Solid Electrolyte Interphase
SEM	Scanning Electron Microscope
SERS	Surface Enhanced Raman Spectroscopy
SICM	Scanning Ion Conductance Microscopy
SiNW	Silicon Nanowire
SKPM	Scanning Kelvin Probe Microscopy
SMM	Scanning Microwave Microscopy
SNOM	Scanning Near-Field Optical Microscopy
SPM	Scanning Probe Microscopy
SSPM	Surface Scanning Probes
SSRM	Scanning Spreading Resistance Microscopy
STAP	Scanning Tunneling Atom Probe
STEM	Scanning Transmission Electron Microscopy
STM	Scanning Tunneling Microscopy
SWNT	Single-walled Carbon Nanotubes
TDS	Thermal Desorption Spectrometry
TEM	Transmission Electron Microscopy
TERS	Tip-enhanced Raman Spectroscopy
THF	Tetrahydrofuran

TR	Torsional Resonance
UHV	Ultra-High Vacuum
UV	Ultraviolet
vdW	Van der Waals
XAFS	X-ray Absorption Fine Structure
XPS	X-ray Photoelectron Spectroscopy
XRD	X-ray Diffraction
μR	Micro-Raman

Abstract

Adoption of sustainable energy technologies requires parallel progress in the field of energy storage. The advances in scanning probe microscopy (SPM) involve properties and size scales of critical relevance to energy-related materials. The design and control of materials at the nanoscale are the foundation of many new strategies for energy generation, storage and efficiency. In the past decade, SPM has evolved into a very large toolbox for the characterization of properties spanning size scales from sub-microns to nanometers. Advanced local probe techniques and in particular emergent scanning probe microscopy (SPM) methods addressing several nanoscale functionalities including electrochemical, near-field optical, electromechanical, thermal probes and combined SPM-scanning transmission electron microscopy have been gradually implemented in energy storage and conversion applications.

On the other hand, friction is an important limitation of energy efficiency performances in electro-mechanical components and it happens that some SPM techniques already used for electric energy storage studies are also the ideal tools to study nanotribological friction. Such friction limitation is particularly crucial in small electric energy components and so this subject assumes a great relevance with the present trend of their miniaturization and integration with micro-electromechanical devices.

The modern tools of nanotechnology, such as scanning probe microscopes and micro/nanomachines, can provide important information about the fundamental nature of space, especially the zero-point electromagnetic field. A better understanding of the force that arises from the zero-point field may enable its control to some extent and so would improve the performance of micro- and nanomachines that utilize the force to achieve contactless transmission or exploit the Casimir ratchet effect.

This book addresses the use of local probes and mainly covers SPM techniques in these areas of research. It is intended as a reference for scientists and as a complement at the advanced graduate level.

Keywords Nanophysics · Nanoenergy · Nanotribology · Nanoprobes · Nanotechnology

Introduction

Nanotechnology is being used in several applications to improve the efficiency of energy generation or develop new methods to generate energy, including renewable energy. Renewable energy sources are an important component of energy strategies, not only to contribute to future energy generation, but also to minimize the environmental impact of energy generation and utilization. Adoption of sustainable energy technologies requires parallel progress in the field of energy storage. The vehicular technology provides a paradigm, since the battery pack cost is a significant fraction of that of the car itself, with weight to match. The commercial viability of electric and hybrid cars hinges on development of high-energy density high lifetime batteries. In parallel to batteries, fuel cell systems, or devices for direct conversion of chemical energy of fuels into electricity are being developed. The intrinsic advantages of fuel cells are very high efficiency approaching thermodynamic limits, and scalability from micro- to mega scales.

Broad range implementation of commercially viable electric and hybrid vehicles is impossible without development of low cost long lifetime and high energy density storage. Similar problems exist for power source for mobile devices (laptop battery has an order of magnitude smaller energy density and many orders of magnitude higher price per watt hour than fossil fuels). Finally, intermittency of solar and wind energy requires sophisticated grid storage technologies, of which batteries can be a significant part.

Electric Energy Storage can not only help in existing or future fossil fuel-based power plants but it can also play a vital role in design and development of power plants based on other unconventional renewable energy sources, such as solar, wind, or nuclear energy [1]. The energy harvested in such power plants varies largely due to operating conditions such as weather, wind speed, and temperature. In this case, Electric Energy Storage (EES) devices can be used to store the energy during the smooth operation of the plant and overcome its intermittent nature. Thus, harvesting energy from unconventional sources would be more feasible. These types of power plants can greatly reduce CO_2 emissions.

Presently, batteries exist in an astonishing variety of shapes, weights, energy, and power densities and compositions, and the originally clear-cut boundaries between fuel cells and batteries are becoming more diffuse, with technologies ranging from regenerative fuel cells to flow batteries to metal–air batteries combining features (both positive and negative) of both approaches.

Classical macroscopic electrochemical techniques provide extensive information at macroscopic level. Only in limited cases do techniques such as Electrochemical Impedance Spectroscopy (EIS) allow individual contributions of surfaces and interfaces to be established with high veracity. Typically, such studies require systematic variation of relative sizes of dissimilar components of the device, chemical conditions, and temperature. However, in all cases information on the lateral homogeneity of properties and its evolution during device operation remain hidden.

The advances in Scanning Probe Microscopy (SPM) involve properties and size scales of critical relevance to energy-related materials. The design and control of materials at the nanoscale are the foundation of many new strategies for energy generation, storage and efficiency. In the past decade SPM has evolved into a very large toolbox for the characterization of properties spanning size scales from sub-microns to nanometers.

Despite the rapidly expanding manufacturing capabilities on the macroscopic devices, the microscopic mechanisms underpinning battery and fuel cell operations in the nanometer-micrometer range are virtually unknown. Such lack of information is due to the dearth of experimental techniques capable of addressing elementary mechanisms involved in battery operation, including electronic and ion transport, vacancy injection, and interfacial reactions, on the nanometer scale. Advanced local probe techniques and in particular emergent SPM methods addressing several nanoscale functionalities including electrochemical, near-field optical, electromechanical, microwave, thermal probes, and combined SPM-STEM (Scanning Transmission Electron Microscopy) have been gradually implemented in energy storage and conversion applications.

At this stage it is also appropriate to draw a few words regarding recent progress in nanophysics for photovoltaic conversion of solar energy. To achieve a pronounced advancement in photovoltaic technology, researchers are currently working on a third generation of cells with nanostructured materials which will achieve efficiencies much higher than the Schockley-Queisser limit. The use of atomic force microscopy with a conductive cantilever has revealed to be an indispensable tool for studying local electronic properties of silicon-nanostructures: p–i–n radial junctions of amorphous Si grown on Si nanowires. They have observed variations of the conductivity of the radial junction solar cells based on Si nanowires. Regarding dye-sensitized solar cells, since nanocrystalline TiO_2 photoelectrode consists of sintered nanoparticles, the microscopic work function and surface photovoltage determination of TiO_2 photoelectrodes using Kelvin probe force microscopy in combination with a tunable illumination system is an appropriate analytical choice.

In order to complete this brief introduction, it is worth to mention that friction is an important limitation of energy efficiency performances in electro-mechanical components and it happens that some SPM techniques already used for electric energy storage studies are also the ideal tools to study nanotribological friction. Such friction limitation is particularly crucial in small electric energy components and so this subject assumes a great relevance with the present trend of their miniaturization and integration with micro-electromechanical devices.

The modern tools of Nanotechnology, such as scanning probe microscopes and micro/nanomachines, can provide important information about the fundamental nature of space, especially the zero-point electromagnetic field. A better understanding of the force that arises from the zero-point field may enable its control to some extent and so would improve the performance of micro- and nanomachines that utilize the force to achieve contactless transmission or exploit the so-called Casimir ratchet effect.

Chapter 1
Local Probes in Energy Storage

Abstract The paradigm of real clean and sustainable energy technologies without having a climate impact will demand new solutions for energy storage. Nanotechnology provides a clue for such endeavor since the advances in scanning probe microscopy (SPM) enable to study properties at the nanoscale which are crucial in energy storage modern technologies. The SPM methods emerged as a tool to address several nanoscale multi-functionalities of energy storage materials as they play a role in rechargeable batteries towards optimization of energy density and lifetime.

Keywords Nanophysics · Nanoenergy · Nanotechnology

1.1 Lab-on-a-Tip and Critical Operation of Energy Devices

Since the invention of Scanning Tunneling Microscopy (STM) in 1981 by the Nobel physicist Heinrich Rohrer, Scanning Probe Microscopes (SPMs) have evolved as powerful imaging devices for obtaining nanometer-resolved topographic information on various objects in vacuum, air, and liquids.

The size scales of the interactions that are critical to the operation of solar cells, fuel cells and batteries are in the range of 1–500 nm, and then scientific and technical advances require local characterization of not only structure but, as important, properties and functionality at these levels. Scanning Probe Microscopy is a platform with the ability to characterize a wide variety of properties, as well as structure, at length scales spanning nanometers to hundreds of microns. This is the information required to obtain the fundamental understanding that will be the basis of knowledge-driven design of energy technologies.

Initial approaches toward extending STM to the study of electrochemical processes involved ex situ preparations followed by imaging in the UHV environment [2–4], and this method complemented other in situ techniques [5, 6]. Later on, the

© The Author(s) 2015
R.F.M. Lobo, *Nanophysics for Energy Efficiency*,
SpringerBriefs in Energy, DOI 10.1007/978-3-319-17007-7_1

1

Sonnenfield and Hansma's studies of the first atomic scale resolution imaging performed in liquid by STM [7] exposed the potential of applying STM to the liquid environment and possible in situ electrochemical analysis. In the following years, researchers examined electrochemical processes using the EC-STM as a structure sensitive probe with interest in performing investigations of reactions at the electrode-electrolyte interface. The attractive nature of in situ imaging results from its intrinsic capability to preserve the electrode-electrolyte interface during imaging. Initially, the ability to perform EC-STM imaging was limited due to a lack of technology in regards to protection of the tunneling tip from interfering faradaic current in the electrolyte environment. The field took giant steps forward when three independent research groups introduced the application of the potentiostatic STM concept [8–10].

One step further came from the pioneering work of Itaya and colleagues through application of a biopotentiostatic circuit toward more dynamic control of the electrolyte environment [9]. The bipotentiostat is capable of independently controlling the substrate and tip potential without affecting the bias between them via a current carrying counter electrode acting to maintain a constant sum of the currents at the tip and the substrate [11]. The tip potential can be held within a potential range of capacitive behavior, thereby minimizing faradaic currents contributions to the measured tunneling current. The nature of the interaction between not only the tip and surface but also the tip and the electrolyte soon became increasingly clarified [12, 13]. In addition to the bipotentiostat, new technology has been developed in the process of tip fabrication and coating as deemed necessary by the susceptibility of both tip and substrate to electrochemical charge-transfer reactions with the electrolyte solution. Reactions of this type have two major effects: (1) random alterations in surface and tip morphology can occur by adsorption or dissolution processes involving the electrolyte; (2) the presence of additional electrochemical currents can drastically affect the STM feedback loop by increasing noise and current fluctuations at the tip. Both of these effects contribute to decreased resolution and reproducibility in EC-STM imaging [14].

Scanning probe microscopy has developed into an ideal toolbox of techniques for interrogating matter at those sub-micron length scales, and the field has dramatically advanced in the last years such that complex properties can now be addressed locally.

The large variety of electrochemical, transport, and mechanical phenomena in energy systems requires a comprehensive approach utilizing many of the SPM variants, and adapting them to reveal energy-related functionalities. Examples of properties that can be probed by SPM that are critical to energy applications include photoconduction, ion dynamics, impedance, and dielectric polarization.

Scanning probes for testing electronic and dielectric properties usually involve a conducting tip and application of an electrical signal to the tip-surface junction. Variants yield, scanning resistance, capacitance, and conductance microscopies, as well as "scanning potentiometry", or still Kelvin probe microscopy.

Another important set of properties in energy materials are related with the dielectric function, which describes the response of a material to electric fields and contains information about atomic polarization, ionic motion, and charge diffusion. The dielectric constant is frequency dependent because these phenomena occur at different rates. Consequently, the frequency dependence of this property probes a range of interactions. Accessing dielectric constant and dielectric function is made possible through capacitance, impedance, and light scattering methods.

A major progress in optimizing energy storage and conversion materials is only possible by the capability to probe locally individual functionalities including electronic and ionic transport and electrochemical reactivity. One approach for exploring local electrochemical functionality is based on scaling down classical electrochemical characterization techniques, by using SPM tip as a ultra-microelectrode. This is achieved by techniques such as electrochemical scanning tunneling microscopy, atomic force microscopy or scanning ion conductance microscopy. In addition, some of classical SPM techniques are incorporated in the in situ battery or fuel cell operation environment to provide information on evolution of surface morphology during the charge-discharge cycles, probe local static strains during the electrochemical processes, locally mapping dc conductive currents and perform SPM-based impedance imaging.

1.2 Aging Studies of Electric Energy Storage Devices by Scanning Probe Microscopy Techniques

Batteries, pseudocapacitors and asymmetric or hybrid electrochemical capacitors are all referred to as electrochemical energy storage devices because their mechanisms of operation involve chemical reactions, unlike the Electric Double Layer Capacitor (EDLC) that does not involve chemical reactions. Despite the term "electric energy storage" can be somewhat suggestive of electric double layer capacitors they will not be focused in this work. Anyway, one can refer that due to the impressive Z resolution (<1 nm) and lateral resolution (~ 10 nm) of the SPM technique, it is used to investigate microporous and mesoporous carbons, demonstrating the effect of pore size on the ion insertion kinetics, and allowing for the anion and cation insertion processes to be separated. During charge/discharge of porous carbon electrodes ions move in and out of the pores resulting in miniscule volume changes. It is then possible to monitor the volume changes in situ under electrochemical control, providing a non-current based method of investigating the kinetics of the insertion/deinsertion of ions into porous carbons. The volume changes of carbons having different surface areas and pore size distributions can be examined to investigate the effect of pore size on electrode strain and ion kinetics in ionic liquid electrolytes. The cation and anion insertion processes can be separated and the kinetics of each examined. This enables to compare both anion and cation kinetics with the carbon pore size.

Nanostructured carbon materials have been widely investigated and utilized for energy storage and conversion devices, especially as electrode materials for supercapacitors because of their high electrical conductivity, high specific surface area, huge corrosion resistance, high temperature stability, and tunable porous structures. In addition, compared to redox electrode of transition metal oxides and conducting polymers, nanostructured carbon electrodes also have merits of longer cycling stability and lower cost.

The aging of supercapacitors is typically caused by impurities introduced during electrode fabrication, especially in the case of supercapacitors utilizing organic electrolytes to provide a higher energy density for stationary applications. In addition to the impurities, the elevated operating voltage and temperature can also accelerate the aging of supercapacitors, which can partially be relieved by selecting a suitable electrolyte solvent.

In order to take profit of the potential of the high power supercapacitors, the surface treatment of the carbon materials, especially carbon nanotubes (CNTs), is very important. The treatment can be used to tune the surface hydrophilicity (in order to increase wettability and specific surface area), to introduce functional groups, and to create surface defect for a better electrolyte/electrode interface formation.

Battery technology is boosting the electronic gadget industries with application in cell phones, laptops, and tablets. In these applications the life of battery usually exceeds the life of the electronic gadget, and so the life of the battery has never been a critical issue. However, in power plants and automobile applications the life of the battery is very critical. Particularly, for successful development of electric vehicles longevity the battery cycle life is essential. It is then crucial to investigate and understand the atomic- and molecular processes that govern the operation of the batteries. Identifying the damage mechanisms in batteries in order to improve their cycle life and longevity urges as a matter of fundamental research.

Batteries and electrochemical capacitors are typical EES devices but fuel cells are not truly EES, although they also convert chemical energy (stored in the source fuel) to usable electrical energy. They differ from others EES in that the reactants flow through the cell rather than being sealed within the cell.

The continuous cycling of a EES device usually leads to its capacity fade and increase in the internal resistance. Thus, capacity and internal resistance are used as the metrics for measuring aging at the system level.

Typical aging mechanisms of the electrodes, common presently in all of those energy storage devices are:

- disordering of crystal structure or phase changes during the extended operating time (ex: intercalation/de-intercalation of Li in the host lattice of Li in the case of Li-ion batteries).
- metal dissolution giving rise to deposits on the electrodes.
- surface films formation due to electrolyte decomposition (ex: oxidation).

In order to deeply understand such mechanisms, a multi-scale characterization route map needs to be implemented. For the specific EES device under study, one needs to create different electrode surfaces samples aged to different percentages of their lifetimes and subsequently performs analysis at the millimeter scale followed by micro- and nano-scale surface characterization. Usually thermography mapping enables to study material degradation on millimeter scale and identify the potential regions (μm^2) of material degradation; this is because it has been measured that thermal diffusivity of aged samples is higher compared to unaged samples.

On its turn, for micro- and nano-scale surface characterization several techniques have to be used to investigate the identified areas further for surface roughness, grain coarsening, chemical and structural changes and surface electrical properties.

Electrochemical Impedance Spectroscopy (EIS) is mostly used for characterization of aged cells. Actually, the direct measure of cell aging is the increase in cell impedance. This increase can be attributed to the increase in surface resistance of the electrodes. The surface resistance affects the battery operation because battery reactions occur at the surfaces of the anodes and cathodes.

Scanning Electron Microscope (SEM) micrographs are usually taken to investigate mechanisms for the increase in thermal diffusivity shown previously with thermography. Very often such micrographs can reveal the coarsening of the nanoparticles in the aged samples, which is believed to be due to sintering. In fact, the coarsening can lead to a decrease in the effective surface area of the particles per unit volume and thus to a decrease in porosity which leads to an increase in the thermal diffusivity.

The investigation of aging at the nanoscale has been implemented in lithium-ion batteries given the advantage of these kind of EES devices, that relies on their high energy and power density. They can be used in stationary applications (conventional or renewable energy power plants) or also in development of electric vehicles. There is a need to understand the aging mechanisms of these batteries to improve their life cycles. Aging of the cells at the macroscopic level is quantified by the change in the internal resistance measured by impedance techniques. The related loss of capacity has its origin on the degradation of the battery electrodes, which material composition deterioration is caused by several complex processes that occur within the batteries.

The main components of the Li-ion battery are the anode, cathode, separator and the electrolyte. The anode in a Li-ion battery is mostly made of in graphitic carbon (on copper current collector), while the cathode is usually nanoparticulate $LiFePO_4$ (on aluminum current collector), and the electrolyte is a lithium salt soluble in an organic solvent. The electrolyte provides the path for the lithium ions to travel between the electrodes during the cycling of the cell. The separator is made up of a polymer that prevents the contact between the anode and the cathode but allows the lithium ions to pass through it. The chemical process is reversible and based on intercalation/deintercalation, i.e. guest ions are introduced in the anode host structure during charging and removed from it during the discharging (Fig. 1.1).

The Li^+ ions are never oxidized in this reaction but the Fe in $LiFePO_4$ undergoes oxidation from Fe^{2+} to Fe^{3+} during charging, and reduction from Fe^{3+} to Fe^{2+} during discharge. This is slightly different than the reaction in conventional batteries where the anode undergoes oxidation and the cathode undergoes the reduction.

In any Li-ion cell with a carbonaceous anode a Solid Electrolyte Interphase (SEI) is formed during cycling of the cell. The SEI layer is formed from the electrolyte decomposition products [15] and prevents the graphite surface from

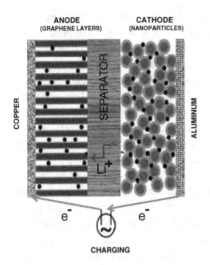 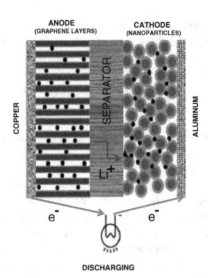

Fig. 1.1 Li-ion cell operation

further exfoliation. It also prevents further reduction of the electrolyte and con-
sumption of active lithium.

Regarding the electrolyte its function is to provide a path for the ions while
blocking any electronic charge flow, and the cycle life of the battery depends on the
exact composition of the electrolyte [16]. The electrolyte should be:

- thermally stable in the operating range of the battery;
- safe and environmentally compatible;
- electrochemically inert with the oxidizing or reducing electrode surfaces. Often
  a passivation film composed from reduction products of the electrolyte is formed
  on the electrode surface, and it affects the cycle life of the battery. The stability
  and quality of the film depends on the electrolyte composition, additives and
  impurities. In general, the electrolyte is composed of one or more liquid aprotic
  solvents and one or more salts.

Thermographic studies of the porous nanoparticulate $LiFePO_4$ cathode have
shown that its thermal diffusivity increases due to aging [17]. Actually, as the
cathode ages, the nanoparticles tend to coarsen by sintering. Due to such sintering,
the effective surface area per unit volume decreases [18], with an associated
decrease in the porosity of the cathode. This decrease in porosity can expose larger
area of the aluminum current collector, and since aluminum has high thermal
diffusivity, the overall thermal diffusivity of the aged sample may show an increase.
Also in general, as the porosity of the medium decreases, the thermal conductivity
increases, and so the aged sample enhances its thermal diffusivity [19].

A brief review of AFM techniques (Fig. 1.2) used in Li-ion battery research has
been provided by Nagpure et al. [21]. The AFM, being a variant of the SPM family

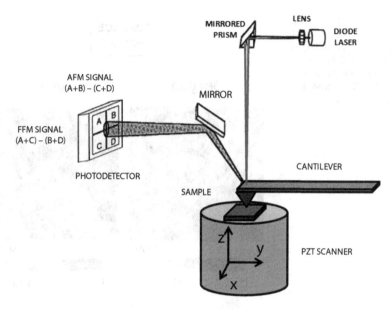

Fig. 1.2 Principle of operation of the atomic force microscope (AFM)

of techniques, provides surface morphology maps of the cathode samples within micron to nanometer resolution. These maps are useful in studying the grain coarsening phenomena. Along with this standard measurement, AFM modules are also helpful in measuring surface electrical properties.

Figure 1.3 displays the AFM surface height image of the LiFePO$_4$ cathode sample collected from the unaged and aged cells. As it can be seen from the images, the LiFePO$_4$ nanoparticles tend to coarsen in the aged samples as compared to the unaged samples. The same phenomena was observed with SEM micrographs at micron length scales. The AFM surface height images have higher spatial resolution than the SEM micrographs, and so they might be more useful to quantify the particle size distribution.

The coarsening of the nanoparticles leads to a decrease in the effective surface area of the particles affecting the rate of the reaction. The change in the particle size would also affect the diffusion kinetics of the lithium ions during charging and discharging cycles. Such coarsening also favors the disbonding of the particles from the aluminum substrate which causes physical failure of the active material and increase in the internal resistance (due to loss of contact). In addition, the coarsening leads to separation of particles and then to loss of contact between them.

Advances in AFM instrumentation has led to the development of a technique known as Scanning Spreading Resistance Microscopy (SSRM) [23]. The SSRM module used in AFM measures the surface resistance between the conductive tip and the sample while the tip is scanned in contact mode across the sample surface. An important application of the SSRM technique is in the mapping of carrier

Fig. 1.3 Surface height maps of a LiFePO₄ cathode sample taken from unaged and aged cells. A +1 V DC bias is applied to the sample and scale: V ∝ 1/R. (reprinted with permission [21])

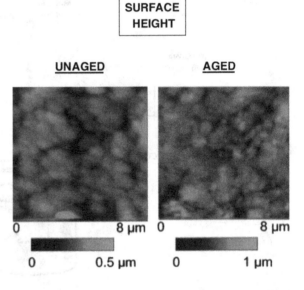

concentration inside a semiconductor device [24]. In contrast to SSRM, there also exists a method called Current Sensing AFM (CSAFM) to measure the surface current between the conductive tip and the sample. This has been previously used in lithium ion battery research. A review of AFM techniques used in electrical characterization of battery materials has been presented in [21].

The nanoscale surface resistance measurements have been performed [21], with a Nanoscope IIIa Dimension™ 3000 AFM equipped with the SSRM application module. The conductive SCM PIT (Veeco Instruments) probes used in such study were coated with Pt-Ir on front and back side and had a nominal tip radius of 20 nm. A known DC bias voltage of +1 V was applied between the sample and the conductive tip, and the current was monitored using a logarithmic current amplifier built into the SSRM sensor, as shown in Fig. 1.4.

Figure 1.5 shows the SSRM surface resistance image of the LiFePO₄ cathode sample taken from the unaged and aged cells. As the module configuration when +1 V is applied between the tip and the sample, the higher voltage reading represent lower surface resistance. The surface resistance scale on the unaged sample is 0–2 V, while the scale on the aged sample is 0–20 mV. The lower voltage output in the case of aged samples indicates higher surface resistance as compared to unaged sample. Thus, surface resistance increases as the cells are aged.

Based on their results Nagpure et al. [21] have proposed a mechanism that leads to the increase in the surface resistance of the LiFePO₄ cathode sample, shown in Fig. 1.6. When the tip scans over the surface of the sample, a circular contact is formed between the LiFePO₄ nanoparticles and the tip. As LiFePO₄ nanoparticles have very poor conductivity [26], in order to increase it, they are coated with carbon [27]. In the case of the unaged sample, the total surface resistance measured is the

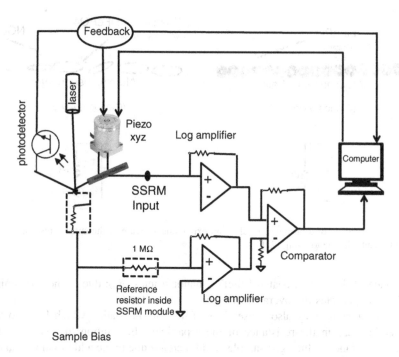

Fig. 1.4 Schematic of the AFM-based resistance measurement technique (SSRM) where the surface height is measured in contact mode, and the resistance is measured by the current resulting from the applied sample DC bias voltage

Fig. 1.5 Surface resistance maps of a LiFePO$_4$ cathode sample gathered from unaged and aged cells. A +1 V DC bias is applied to the sample and scale: V ∝ 1/R. The lower voltage output in the case of aged sample indicates higher surface resistance as compared to unaged sample. (reprinted with permission [21])

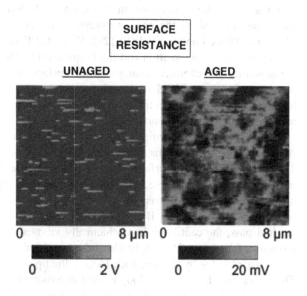

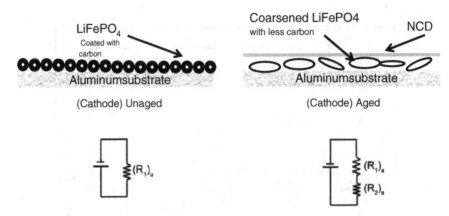

Fig. 1.6 Schematic of a proposed mechanism explaining increase in the surface resistance of a LiFePO$_4$ cathode due to aging [21]

resistance of the carbon-coated LiFePO$_4$ nanoparticles (R_u); due to the coarsening of the nanoparticles the overall resistance increases to (R_a).

The coarsening may also causes loss of the carbon coating which leads to the further increase in the resistance of these particles. In addition to this, the total surface resistance of the aged sample could increase due to the additional resistance from the Nano-crystalline Deposit (NCD) formed on the cathode surface. NCD is formed due to the chemical reactions taking place at the surface of the cathode.

The AFM Kelvin Probe Microscopy (KPM) technique, which has been used in a variety of nanotribological applications to measure surface potential [20], also found a successful application in the study of Li-ion batteries [28]. The KPM technique is based on the contact potential difference method for measuring the Electronic Work Function (EWF) [29]. Since EWF is strongly influenced by the surface chemical composition and the Fermi level of the material, KPM can detect the structural and chemical changes of the surface and provide information about the onset of damage. KPM was used to characterize aging of the cathode surfaces by measuring the change in the surface potential which can be attributed to physical and chemical changes of surface.

Nanoscale surface potential measurements were taken with a Dimension™ 3100 AFM. A schematic of this instrument with KPM setup is shown in Fig. 1.7. The KPM measures the surface potential of the samples in interleave mode. Along one scan line on the sample, in first pass, the surface height image is obtained in tapping mode. In second pass the tip is lifted off the sample surface and a surface potential map is obtained. Both images are obtained simultaneously [30]. During the first pass, the cantilever is mechanically vibrated by the X-Y-Z piezo near its resonance frequency. The amplitude of the tip vibrations (not shown) is maintained at a constant value by the feedback loop as the tip scans the surface of the sample. The signal from the feedback loop is used to construct the height map of the sample surface (Fig. 1.7a). During the interleaved scan, the X-Y-Z piezo is switched off.

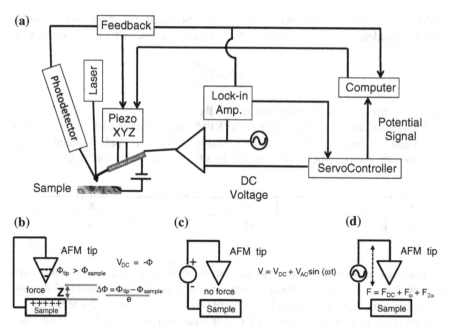

Fig. 1.7 a Schematic of the two pass interleave scan method used in KPM. (Adapted from [30]). **b** Electrostatic potential and interaction force between a conducting tip and a sample (for illustration $\Phi_{tip}\Phi_{sample}$ is assumed), **c** external DC voltage applied to nullify the force, and **d** external AC voltage with adjustable DC offset is applied to the tip which leads to its vibration (Adapted from [31])

Instead, an AC signal is applied directly to the conductive tip which generates an oscillating electrostatic force on the tip. The tip is scanned along the surface topography line obtained in the tapping mode with a certain lift off the sample (dotted line in Fig. 1.7).

To briefly describe the operating principle of KPM, consider a tip and sample interaction as seen in Fig. 1.7a–d. When the tip and sample are electrically connected (Fig. 1.7a) electrons flow from the material with the lower work function to the material with the higher work function. Due to the difference in the work function of the electrically connected tip and the sample an electrostatic contact potential difference (or surface potential difference) is created between the tip and the sample [32]. The value of this surface potential (Φ) is given by:

$$\Delta\emptyset = \frac{\emptyset_{tip} - \emptyset_{sample}}{e} \tag{1.1}$$

where Φ_{tip} and Φ_{sample} are work functions of the tip and the sample, respectively, and e is the electron's charge. Φ will be affected by any adsorption layer and the phase of the material near the surface. Electrostatic force is created between the tip and sample under the influence of this surface potential difference and the separation dependent local capacitance C of the tip and sample. This force is given by

$$F = \frac{1}{2}(\Delta\emptyset)^2\frac{\partial C}{\partial z}$$

(1.2)

where z is the distance between the tip and the sample.

Along with $\Delta\Phi$, in the operation of the KPM a compensating DC voltage signal (V_{DC}) and AC voltage signal, $V_{DC}\sin(\omega t)$ (Fig. 1.7c, d), is applied directly to the tip. Thus the electrostatic force between the tip and the sample becomes:

$$F = \overbrace{\frac{1}{2}\frac{\partial C}{\partial z}\left\{(\Delta\emptyset + V_{DC})^2 + \frac{V_{AC}^2}{2}\right\}}^{DC_{term}} + \overbrace{\frac{\partial C}{\partial z}(\Delta\emptyset + V_{DC})V_{AC}\sin(\omega t)}^{\omega_{term}} - \overbrace{\frac{1}{4}\frac{\partial C}{\partial z}V_{AC}^2\cos(2\omega t)}^{2\omega_{term}}$$

(1.3)

The cantilever responds only to the forces at or very near its resonance frequency. Thus, only the oscillating electric force at ω acts as a sinusoidal driving force that can excite oscillations in the cantilever. The DC and the 2ω terms do not cause any significant oscillations of the cantilever. In tapping mode, the cantilever response (RMS amplitude) is directly proportional to the drive amplitude of the tapping piezo.

In the interleave mode the response is directly proportional to the amplitude of the term [30]. The servo controller applies a DC voltage signal equal and out of phase with so that the amplitude of the tip becomes zero $(F = 0)$. This compensating signal from the servo controller creates the surface potential map of the sample [31].

A conductive AFM tip is necessary for the KPM experiments. The tips used in these experiments had an electrically conductive 5 nm thick chromium coating and 25 nm thick platinum coating on both sides of the cantilever (Budget Sensors, Model Multi75E-G). The resonant frequency of the tips was 75 kHz, and the radius was less than 25 nm. The interleave height was optimized to 150 nm for a good surface potential signal.

The surface potential maps of the unaged and aged LiFePO$_4$ cathode samples are shown in Fig. 1.8. The maps were collected by applying +1.0 V and +3.3 V from the external DC voltage source. Maps for samples without an externally applied voltage are shown for comparison. Within each surface potential map for both the unaged and aged sample, one observes no large difference in the contrast. This suggests an almost uniform dissipation of charge over the surface of the samples under the externally applied voltage. Thus indicating that under the external source the sample tends to charge uniformly even in an aged condition. The uniform charging is good for the cell because it avoids large local currents and subsequent damage to the cathode.

From the normal distributions one can conclude that the mean values of surface potential on the aged sample are lower than that of the unaged sample. Thus, even though the externally applied voltage was the same for the unaged and aged

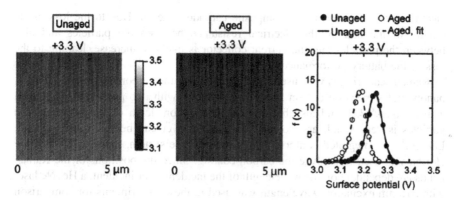

Fig. 1.8 Surface potential maps of the unaged (*left column*) and the aged (*middle column*) LiFePO$_4$ cathode samples with an external voltage of +3.3 V. The *right column* shows the normal probability density distribution of the surface potential values obtained for the unaged and aged samples. The mean value of the surface potential decreases after aging. (reprinted with permission [22])

samples, the charge sustained on the surface of the age cathode is less than that sustained on the unaged cathode.

The surface potential measured with KPM is the difference between the work functions of the tip and sample surface. Since the same kind of tip is used in these experiments, the work function of the tip is constant in each surface potential map. The work function is the property of the structure near the surface of the sample along with the chemical potential of the surface. The decreased surface potential in the aged sample is the indication of the surface modification and can occur due several factors. A phase change of the cathode material occurs from LiFePO$_4$ to FePO$_4$ and back to LiFePO$_4$ during respective charging and discharging cycles of the battery. During charging, the Li ions from the LiFePO$_4$ cathode are intercalated in the graphite anode. During discharging, the Li ions move out of the graphite anode and are intercalated back in the cathode. The structured LiFePO$_4$ has a different work function from that of the metastable FePO$_4$. The change in the surface potential map of the aged sample indicates that one of the phases might be growing in the cathode sample during the battery aging.

In summary, KPM is used to measure the change in the surface potential. The aged sample showed less charge sustaining capacity as compared to the unaged sample. The loss in the ability of the cathode to retain the applied charge can also be attributed to the coarsening phenomena. In this case, due to the larger particle size the overall distance travelled by the applied charge is increased thus leading to the loss of capability to sustain the applied charge within the applied time.

Regarding multi-scale chemical and structural characterization studies of the battery cathode aging (using several analytical techniques applied to the LiFePO$_4$ nanoparticles), changes in the lithium concentration, local lithium bonding and local Li environment, have been revealed [22]. For instance, Raman studies have shown degradation in the quality of the carbon coating. This has direct effect on the electronic conductivity of the composite cathode material. Due to the loss in the quality of the carbon, the electrical resistance between the particles and also between the particles and the current collector is likely to increase, leading to the loss of the battery performance.

Raman scattering is very useful in characterizing the carbon coating of $LiFePO_4$ nanoparticles, because carbon is a strong scatterer with two predicted E_{2g} Raman active modes [33], and also because the penetration depth of light inside those particles is very small, and so the first coating layer can be easily probed [34]. Labram®, an integrated confocal Raman microscope system, made by ISA Group Horiba was used to analyze the carbon-coating. Since the positions of the Raman bands are dependent on the wavelength of the incident laser incident, a He–Ne laser with 512 nm excitation wavelength was used in these experiments for comparison of the data with the published literature. The laser power was adjusted to 2.5 mW and the laser spot size was at 5 μm. RS experiments were conducted under ambient conditions at room temperature.

Doeff et al. [35] have shown that electrochemical performance of carbon coated $LiFePO_4$ cathode is not only dependent on the quantity of the carbon but also on the quality of carbon. According to the Raman studies the carbon coating degrades in quality as the batteries are aged at higher C-rates. The higher intensity ratio observed for the two main peaks in the Raman spectrum (in the case on batteries cycled at higher C-rate), indicate poor quality of carbon leading to loss of electrical conductivity and subsequent decrease in the performance of the battery.

1.3 Scanning Probe Techniques in Energy Storage and Harvesting

Critical to the advancement of kinetic models in fuel cells involving oxygen reduction and fuel oxidation processes has been the parallel development of complimentary in situ measurement techniques. Cell performance is typically collected with either voltammetry or Electrochemical Impedance Spectroscopy (EIS), yielding relevant measurements such as those of power density and impedance. While these techniques have demonstrated effectiveness in determining overall performance under various operating conditions (i.e. fuel, temperature, microstructure), correlating measured results with the underlying mechanisms that govern the electrochemical conversion of fuel into energy is rather difficult.

Various spectroscopic methods are typically used in tandem to correlate compositional and phase changes with variations in the performance metrics. These techniques include X-ray Photoelectron Spectroscopy (XPS) and X-ray Absorption Fine Structure (XAFS), as well as non-destructive optical methods such as Raman and IR spectroscopies, which require the engineering of optical access while preserving cell integrity. Recent progress in performing SPM studies at elevated temperature ranges offer a platform to complement existing spectroscopic methods in accessing direct information of physical processes at the nanoscale.

The need for force-based electrochemical SPM methods is due to the lack of local information on battery functionality that can be gained through traditional characterization techniques. Conventional electrochemical methods employed to characterize battery materials largely utilize current sensing detection (ex: Galvanostatic Cycling (GC), Cyclic Voltammetry (CV), Electrochemical Impedance Spectroscopy (EIS)), all of which adequately describe the bulk characteristics of the electrodes and electrolyte. However, as particle size is within the nanometer regime, novel measurement protocols are necessary to describe the behavior of individual particles. For example, most current amplifiers or potentiostat/galvanostats exhibit current limitations on the order of 10 pA. This constraint (10 pA for 1 s) would correspond to a spherical Li particle with a diameter of 137 nm, well above the size considered acceptable for high surface area battery and fuel cell materials. Conversely, modern AFM instruments are capable of measuring surface displacement on the sub-nanometer scale, which represents several orders of magnitude lower than current-based detection schemes. In addition, current-based detection only measures global properties of the material and hence local grain-scale behavior is not possible to be revealed. Once connecting directly to individual grains via current collectors is unrealistic at this short length scale, force-based SPM approaches clearly offer a promise for exploring electro-chemical phenomena on the nanoscale.

Conducting-probe Atomic Force Microscopy (CP-AFM) is a variation of AFM and STM where an electrical current is monitored simultaneously with topography (height). This enables the correlation of spatial features on the sample with their conductivity, and allows to differentiate CP-AFM from STM (where only current is measured). As in normal contact mode AFM, a cantilever affixed with a sharp tip is placed in contact with the surface to be probed, and the deflection of the cantilever is monitored by an optical lever and position sensitive photodiode (Fig. 1.9). The deflection signal from the photodetector is used to control a feedback loop, which uses a piezoelectric actuator (scanner) to move the tip up or down relative to the surface (z direction), maintaining a constant cantilever deflection and thus a constant tip-surface force. The piezoelectric actuator is also used to position and raster the tip across the surface in the x-y plane while the z displacement signal is used to generate a topographic map of the surface.

In CP-AFM, the tip is either constructed of, or coated with, an electronically conductive material and connected to a high gain current amplifier. CP-AFM can be operated in two principle modes; imaging mode and spectroscopic mode. In imaging mode, the conductive tip is scanned over a small sample area while voltage bias is applied between the tip and the sample, generating simultaneous topographic and current images. The polarity and magnitude of the bias can be chosen to achieve more sensitivity. In spectroscopic mode the tip is held stationary while the voltage bias is swept. This enables recording conventional current-voltage (IV) curves from small areas of the sample, and then allows for extraction of quantitative information on the local electronic properties, such as local density of

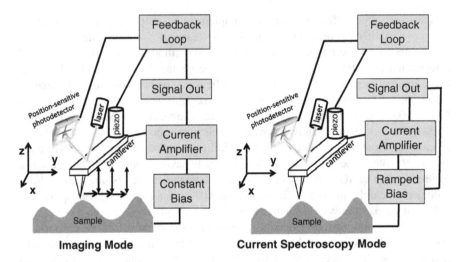

Imaging Mode   **Current Spectroscopy Mode**

Fig. 1.9 Schematic of a conducting-probe atomic force microscope operating in imaging mode (*left*) and current spectroscopy mode (*right*). In both modes an optical lever and position sensitive photodiode measure the deflection of a sharp-tipped cantilever in contact with the sample. A feedback loop connected to a piezo actuator maintains a constant cantilever deflection. In imaging mode, a constant bias is held between the tip and the sample while the piezo actuator scans the tip across the sample's surface. The resulting feedback loop signal and measured currents generate corresponding topography and electrical current maps. In spectroscopy mode, the tip is held in one spot on the surface and the bias between the tip and sample is ramped. This generates a current-voltage curve for the local region of the sample contacting the tip

states and tunneling barrier heights. In selected cases, both imaging and spectroscopy modes can be implemented to perform area-selective nano-fabrication where the tip is used to modify local surface chemistry (i.e. chemical oxidation, etching and electroless plating). The z feedback loop can also be controlled by intermittent contact (tapping) mode or Torsional Resonance mode (TR mode). In these modes, the cantilever is driven by piezo actuators to oscillate vertically along its main axis (tapping) or by twisting about its main axis (TR mode). Interaction with the sample surface dampens the oscillations, and the feedback loop moves the cantilever in the z direction to keep the damping at a set level. Use of these feedback modes is gentler on both the sample and tip, greatly decreasing the lateral forces applied.

As with any AFM technique, CP-AFM requires judicious choice of cantilevers to avoid sample damage and the generation of image artifacts. For instance, the Van der Waals (vdW) molecular solids and polymers imaged with this technique are easily damaged and require low contact forces (thus low spring constant cantilevers must be used). The metal coatings on cantilevers are generally only a few tens of nanometers thick and are easily removed in the contact region by the passing of sustained high current densities or abrasion by hard materials (such as metal oxides). Cantilevers missing the conducting film at their tip can generate current image artifacts. Since current only flows when the sides of the tip contact adjacent topographic features, flat regions will appear non-conducting and regions with high topographic contrast will show conductivity.

CP-AFM is indeed a powerful tool for characterizing local structure and electronic properties. It has been used to investigate the properties of electrodes and battery materials, as well as semi-conductor interfaces.

Electrochemical Strain Microscopy (ESM) is another recent technique which is able to probe Li insertion and diffusion in battery and electrochromic materials, on a finer scale than even spectro-electrochemical microscopy techniques. The detection principle is based in local strain response to change in ionic concentration induced by probe external bias. ESM is based on the intrinsic link between the molar volume of material and Li (or other ion) concentration. Ionic currents, unlike electronic currents, are typically associated with strains on the order of 10–100 % during the lithiation-delithiation cycle. Combined with the huge height sensitivity of dynamic SPM, the detection of local strain generated by bias-induced ionic redistribution potentially allows the measurement of minute changes in Li-ion concentrations. In ESM, the SPM tip concentrates a periodic electric field in a nanoscale volume of material, resulting in local Li-ion redistribution. Induced changes in molar volume cause local surface expansion or contraction (strain) that is transferred to the SPM probe and detected by the microscope electronics. Consequently, this strain-based detection allows high-resolution nanoscale mapping of Li-ion dynamics, and provides previously unobserved details of ionic flows in the electrical storage materials. By combining local strain detection with time-, frequency-, and voltage-spectroscopies, classical electrochemical methods can conveniently be performed on nanometer-scale regions of materials.

In the simplest form of ESM imaging, a small amplitude and high frequency bias are applied to one battery electrode. The ESM imaging mode is based on detecting the strain response of a material to a locally or globally applied electric field through a blocking or electrochemically active (functionalized directly, or placed in an ion-containing medium) SPM tip. The application of high frequency (0.1–1 MHz) electrical excitation to an SPM tip results in Li redistribution and associated changes in the local molar volume, and the resulting surface strain can be detected by the AFM probe. For typical lithium diffusion coefficients ($10^{-14} - 10^{-12}$ m^2 s^{-1}) and characteristic probing depth (10–100 nm), the diffusion time of Li$^+$ (0.1–1 s) is well above the period of potential oscillations (10^{-5} s). This way, the oscillatory response results from very small surface changes of Li concentration. However, the strains generated during galvanostatic cycling studies are much larger than detectable values in piezoelectric materials, supporting the applicability of the ESM approach. Furthermore, these weak changes (as opposed to full charge-discharge cycle probing employed in previous SPM studies) allow high-resolution imaging while maintaining the reversibility of the process, even for large-voltage amplitudes. The remarkable aspect of the strain detection in ESM, as opposed to current-based techniques, is that the signal originates only from ion-motion induced strains, whereas electric currents may contain contributions from electronic conduction (dc) and double-layer and instrumental capacitances.

In brief, the advantages of ESM are the following:

- no range of materials restrictions
- spatial resolution below 10 nm
- ionic current sensitivity below 100 ions
- possible to work in three spectroscopic modes (voltage, frequency and time).
- allows chemical identification if combined with optical methods
- ESM is competitive with Near-Field Spectroscopy (NFS) like SNOM-Raman but has lower costs and is more user friendly.

Scanning Near-Field Optical Microscopy (SNOM) is an optical near-field scanning probe technique and with NFS all the information is collected by spectroscopic means instead of imaging in the near field regime; consequently one can probe spectroscopically with sub-wavelength resolution. In brief, some of the common near-field spectroscopic techniques are:

i. Direct local SNOM-Raman (R-SNOM), where apertureless SNOM can be used to achieve high Raman scattering efficiency factors (around 40). Topological artifacts make it hard to implement this technique for rough surfaces.
ii. Tip-enhanced Raman Spectroscopy (TERS) is a branch of Surface Enhanced Raman Spectroscopy (SERS). This technique can be used in an apertureless shear-force SNOM setup, or by using an AFM tip coated with gold. The Raman signal is significantly enhanced under the AFM tip. A highly sensitive opto-acoustic spectrometer is commonly used for the detection of the Raman signal.
iii. Fluorescence SNOM (F-SNOM) makes use of the fluorescence for near field imaging, and uses the apertureless back to the fiber emission in constant shear force mode. This technique uses merocyanine based dyes embedded in an appropriate resin and edge filters for removal of all primary laser light. This way, resolution as low as 10 nm can be achieved.
iv. Near-field infrared spectrometry (NFIS) and near-field dielectric microscopy (NFDM).

The ability to construct very thin photovoltaic absorbers (tens to hundreds of nanometres thick) is a clue towards high-efficiency photovoltaic device designs that may absorb the full solar spectrum. This might be possible by using light trapping through the resonant scattering and concentration of light in arrays of metal nanoparticles, or by coupling light into surface plasmon polaritons and photonic modes that propagate in the plane of the semiconductor layer. Nanoparticles with sizes lower or comparable to wavelength of sunlight can interact strongly with incident photons, producing a strong scattered radiation and generating high electric fields intensity in the photovoltaic medium.

Ideally, antenna concepts developed at radio frequencies can directly be scaled down to the optical regime because of the frequency invariance of Maxwell's equations. This downscaling implies technological challenges of nanoscale antenna fabrication and the optical properties of metals imply increased absorption losses.

Fortunately, the antenna performance can be enhanced by plasmon resonances that lead to strong and confined fields within the dielectric sub-wavelength gap between two resonant metal wire antennas.

In principle, nano-antenna arrays can be designed to absorb any frequency of light and its resonant frequency can be selected by varying its length. This an advantage over semiconductor photovoltaics, because in order to absorb different wavelengths of light, different band gaps are needed.

It is the local-field distribution at the antenna that determines its functionality, and different techniques have been used to study the spatial confinement and field enhancement around nano-antennas and in antenna gaps. Nano-antennas have been imaged with scattering-type SNOM [36, 37] and the imaging of the local-field distribution is in fair agreement with theoretical predictions. It becomes clear that the SNOM probe itself can interfere with the intrinsic properties of the structure under study, even affecting the resonance conditions. Therefore, approaches free of scanning tips are preferred, and single molecules can be used to probe the local field of a resonant antenna [38].

SNOM exploits near fields, which are non-propagating electromagnetic fields that exist only in the vicinity of their source. Near-field optical probes represent the key components for the performance of SNOM. Basically there are two main classes of sensors: aperture and apertureless probes. Particular, attention has been focused on optical fiber probes and on nano-antenna probes.

Usually, the SNOM probe is placed close to the object being imaged and both the probe and the object are illuminated by light. The near-field interaction between the two leads to an electromagnetic response that is very different from their individual responses in isolation. Among the different types of SNOM probes, plasmonic nano-antennas are of great interest because of their ability to efficiently couple the non-propagating near fields and the freely propagating light from the illumination source. Theoretically a metallic carbon nanotubes (CNT) can act as nano-antenna with a resonance frequency that varies from the terahertz (THz) to the near-IR (NIR) regimes, depending on the length of the CNT.

Fullerene nanoclusters are good candidates for designing miniaturized electrodes with high-surface area and robust films on desired electrode surfaces with a well-controlled morphology. Nanostructuring of fullerene C_{60} beam deposited films is achieved by electrochemical reduction in a potassium hydroxide aqueous solution. Alkali fulleride clusters are formed at the electrode, as it is illustrated by cyclic voltammetry, X-ray diffraction, and scanning tunneling microscopy [39]. Fluorescence emission from fluorophore doped fullerene reduced films was investigated by fluorescence spectroscopy and scanning near-field optical microscopy [39]. These techniques lead to results which fit nanometer-sized fulleride cluster interpretation. In particular, the fluorophore fluorescence lifetime decreases as long as aggregation in the film is more effective, which occurs with the increase of film thickness [39].

In organic solar cell developments, another SPM technique called Electrical Scanning Probe Microscopy (ESPM), play an important role. Actually, nanoscale film morphology is critical since there is an inherent length mismatch between the

10 nm exciton diffusion length in organic semiconductors and the 150–200 nm thickness required to absorb a significant proportion of incoming light. To account for both of these issues, the active layer in organic devices is often a bulk heterojunction, wherein the electron donor and electron acceptor materials are co-deposited onto the substrate to form an active layer of thickness 100–200 nm, with interpenetrating nanoscale domains of donor and acceptor. Though Scanning Tunneling Microscopy (STM) use is widespread in the semiconductor literature, its use in solar cells is limited due to the conductivity variations in typical organic bulk heterojunctions.

Conductive Atomic Force Microscopy (cAFM) and Photoconductive Atomic Force Microscopy (pcAFM) are contact-mode techniques, while Electric Force Microscopy (EFM) and Scanning Kelvin Probe Microscopy (SKPM) are non-contact mode techniques. The cAFM and pcAFM simultaneously records topography and electrical current information. Unlike STM, which uses tunneling current, cAFM uses force feedback mechanism.

On its turn, depending on the feedback mechanism, EFM measures variations in the electrostatic force and/or force gradient that arise from local differences in chemical potential and/or dipole moment. In the simplest EFM version, a constant voltage is applied between the sample and a conductive cantilever and the total shift in resonance frequency at each point above the sample is recorded. Typically, the shift is recorded while the cantilever is retracted from the topography during the retrace in a two-pass scan.

Usually, the resolution of EFM is in the tens of nanometers and interpretation of EFM data can be made problematic due to issues of the cantilever and beam effects, tip shape, and scan height. Beam effects are reduced in gradient detection, but must be considered nonetheless.

The data extracted from EFM can be used to quantitatively map charge trapping within a semiconductor layer. By measuring the electrostatic potential as a function of position, then taking the derivative to find the electric field, it is possible to make direct tests of different models of carrier injection at the metal/semiconductor interfaces in organic field-effect transistors.

EFM has also been used on solar cells like CdTe/CdS, GaInP and GaInP/GaAs/Ge devices to identify the n-p and triple junctions present and to study carrier transport in systems of relevance to solar energy harvesting like PbSe quantum dot arrays. EFM is also particularly valuable because it can yield information about trap character and defect density in devices.

Scanning Kelvin Probe Microscopy (SKPM) is one of the most common electrical scanning probe methods, and the frequency shift mode is generally viewed as more robust. SKPM provides a measure of the Contact Potential Difference (CPD) between a reference probe and the substrate, measured by nulling out the force or force gradient signal with a feedback loop that adjusts a tip-sample bias to null out the CPD. Thus, SKPM is a relative measurement. If one has a reference sample with a known work function, then it is possible to use SKPM to measure relative local work function variations in an unknown sample.

When two SKPM images are taken, one in the dark and one under constant illumination, the difference provides a surface photovoltage image. This technique allows to measure properties such as diffusion lengths and band bending in bulk solar cell materials. Therefore, SKPM combined with optical excitation allows to map these properties with nanoscale resolution.

The so-called Surface Scanning Probes (SSPM) enable to extend surface potential/Kelvin probe microscopy. In SSPM when the tip is above the surface and a static and a periodic field are applied, the resulting force at the tip can be detected via the motion of the cantilever. The first harmonic will be zero when the tip and sample have the same potential. Tuning a dc bias until the force is zero yields the sample potential The force at the second harmonic of applied bias is related only to the capacitance. So the capacitance can be isolated by detection at the second harmonic. In the ideal case of a flat surface, this would yield the dielectric constant at the frequency of the oscillation.

Regarding a Fuel Cell (FC), the key limiting factors that determine its overall performance and hence its commercialization, are the cathode limitation and the Oxygen Reduction Reaction (ORR). The efficiency of fuel cells is primarily dominated by the slow kinetics associated with electrocatalytic oxygen reduction reaction and its activation. This typically requires the use of higher electrocatalytic loadings thereby increasing the cost of operating FCs. Large overpotentials which limit efficiency also increase potential for stray reactions and deterioration of cathode, thus leading to lower lifetimes. Therefore, a better understanding of the factors influencing the kinetics of this reaction at a micro- level is likely to contribute for optimizing the lifetimes and efficiencies of FC devices. Such optimization of FC technology requires the capability to probe structure and functionality of the materials on the whole range of length scales, from atomic to device level. While structural probing is accessible to the present variety of electron microscopy methods, local electrochemical properties required to understand energy and power densities, life times, and degradation remain elusive. To make progress in understanding and optimizing energy conversion and storage materials it is indispensable to have the capability to probe oxygen vacancy and proton diffusion and electrochemistry of gas-solid and gas-liquid reactions.

In parallel, there is a big gap between electrochemical techniques (macroscopic level) and atomic-level transport processes suitable to scattering probes and first principle calculations. Electrochemical characterization of FCs on the macroscopic scale is usually performed using current-based measurements, which are not scalable down to nanometer levels. As an example, the use of micro-patterned electrodes to probe solid oxide fuel cell functionality is limited, while the length scale from micron to nanometer remains generally inaccessible. Such drawback of nanoscale electrochemical characterization capabilities is linked to the difficulties in detecting Faradaic currents due to electrochemical process in small volumes. In order to study the electrochemical activity at the nanoscale, the detected currents will have to be well below the detection limit of most current voltage amplifiers. Local electrochemical probing is then viable when tip acts as electrocatalyst activating local reaction, although applicability of current-based techniques for probing

ORR reactions on the nanoscale is limited. Actually, the techniques based on impedance measurements by SPM can provide information on local impedances of grain structures.

The Electrochemical Strain Microscopy (ESM) is a suitable developed technique for probing electrochemical reactivity and ionic flows in solids down to 10 nm, thus extending the capability of existing SPMs from probing electronic currents and forces to also probing ionic currents. This microscopic tool is based on the detection of electrochemical strain associated with ionic and vacancy movement during bias induced oxygen reduction/evolution reactions, providing a detection strategy alternative to electronic (Faradaic and conductivity) current based strategies. In ESM, a biased SPM tip concentrates an electric field in a nanometer-scale volume of material, inducing an interfacial electrochemical process at the tip-surface junction and diffusive/electromigrative ionic transport through the solid. The confined electric field results gives rise to the injection/annihilation of oxygen vacancies and subsequent vacancy transport and migration induced by the bias. The intrinsic link between the concentration of ionic species and molar volume of the material results in the electrochemical strain and surface displacement. Typically, the ESM uses a differential detection method (in which 2–5 pm surface displacements can be measured at frequencies 0.1–1 MHz) based in the conventional SPM optical beam deflection system. This way, such high-frequency electrochemical strain signal constitutes the basis of ESM detection. This strain detection approach enables probing volumes one hundred times smaller than accessible through current-based electrochemical methods.

In Fig. 1.10 it is illustrated ESM probing method for solid oxide fuel cell materials. The SPM tip is brought into contact with the surface and the electrochemical potential of mobile oxygen vacancies at the tip-surface junction compared to the bulk shifts due to the applied electric field. At low biases, electronic and ionic flows are viable in the material below the tip. Due to the extremely localized character of the SPM probe, application of even weak biases results in high local

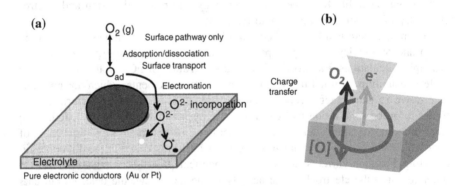

Fig. 1.10 a Schematic representation of ORR/OER reaction mechanism in pure conductors and mixed-ionic electronic electrodes, **b** ESM approach for probing local ORR/OER activity. (Part (**b**) Reprinted from [47])

fields (e.g. 100 mV applied at 10 nm tip-surface contact area yields 10^7 V/m). Consequently, tip-induced electrochemical phenomena can be observed at much lower temperatures as is the case for macroscopic analogs. At sufficiently high probe bias, the potential drop at the junction can activate the ORR process, resulting in generation or annihilation of vacancies depending on the sign of the bias. Under the combined effect of the electric field and concentration gradient, the vacancies diffuse and migrate through the material, resulting in associated changes in molar volume and electrochemical strains. In fact, the associated dynamic surface deformation detected by a SPM can reach the 2–5 pm level.

It must be stressed that while current sensitivity in SPM is subject to the same limitations as macroscopic electrochemical techniques, it offers outstanding sensitivity in detection of surface deformations. Thus, the technique enables detection of local electrochemical activity at the nanoscale which is otherwise improbable using current detection based techniques. It should also be noted that in ESM, the tip plays the dual role of mobile electrode and electrocatalytic nanoparticle, thus allowing for systematic comparison of electrocatalytic activity between different combination of tip materials and local regions of the surface. Using electrochemical strain detection, direct measurements of ORR on the nanometer scale (in volumes around 10 orders of magnitude smaller than possible by conventional electrochemical methods) can be demonstrated.

Exploring combinations with other local techniques such as micro-Raman (μR) or SNOM, variable atmosphere measurements, and ex situ local analysis of reaction regions can establish the exact chemical origins of observed phenomena. The nanoscale probing of the ORR kinetics and oxygen vacancy diffusion will provide a step forward in the ability to control the mechanisms behind the efficiency of air-based fuel cells and metal-air batteries. In particular, the capability to directly link local structure and electrochemical activity is paving the way for the success of systematic electrocatalysis studies and establishing the bridge between atomistic theory and macroscopic electrochemical measurements. Together, this will allow for knowledge-driven design and optimization of energy conversion and storage systems.

Due to their particular thermal properties, carbon nanostructures and in particular carbon nanotubes are suitable candidates for incorporating thermal energy storage devices with high energy density. One of the hindrances which need to be overcome is their very low pristine solubility in water. CNTs tend to form bundles in solution, which is one of the obstacles for its application. Therefore, there is great demand to effectively solubilize CNTs in order to realize wide application of this interesting macromolecule. Water is the most friendly chemical solvent and so in the last years several attempts have been tested in order to meet such objective, which will make easier further chemical stuff synthesis.

One successful water solubilization procedure of carbon nanoparticles, single walled nanotubes (SWCNTs), nano-onions and nanodiamonds has been achieved through their covalent functionalization by fluorination and subsequent derivatization with sucrose [40]. The AFM images provide direct evidences of covalent functionalization of SWCNTs by revealing coating on the backbones of nanotubes.

From the cross section height analysis the size of the individual nanotubes with sidewall attached molecules (~ 2.6 nm) reasonably agrees with the sum of the average F-SWCNT diameter (about 1.3 nm) and approximate size of a sucrose molecule (~ 1.0–1.3 nm) [40].

Still in the aim of energy storage in nanostructured materials with high energy density, a recently developed thermal desorption spectrometry variant, called Molecular Beam-Thermal Desorption Spectrometry (MB-TDS) is used, to monitor in real time and in situ the dehydriding kinetics from hydrogen storage materials [41]. Renewable hydrogen is one clue for the transition from the present carbon based fuels to a free-carbon energy technology, and hydrogen storage at room temperature into a solid matrix is indeed a goal to pursue.

MB-TDS represents a real improvement in the accurate determination of hydrogen mass absorbed into a solid sample. The procedure allows to study dehydriding kinetics at the nanoscale in vacuum, even when the amount of hydrogen is below the detection limit of a microbalance [41]. Among the most efficient techniques for hydrogen desorption monitoring, thermal desorption mass spectrometry (TDS) is a very sensitive one, but in certain cases can give rise to uptake misleading results due to residual hydrogen partial pressure background variations. On the contrary, the variant MB-TDS based on the effusive molecular beam technique represents a significant improvement in the accurate determination of lowermost amounts of hydrogen absorbed on a solid sample [42]. The enhancement in the signal-to-noise ratio for trace hydrogen is on the order of 20 %, and no previous calibration with a chemical standard is required [43]. MB-TDS analysis for lowermost amounts of solid hydrogen contents also has advantages over Atom Probe Mass Spectrometry (APMS). Actually, APMS presents limits to physical shape of samples and involves laborious methods of sample preparation in UHV [44].

Solid oxide fuel cells offer the most efficient and cost-effective means for utilization a FC of a wide variety of fuels beyond hydrogen. Their performance and the rates of many chemical and energy transformation processes in energy storage and conversion devices in general are limited by charge and mass transfer along electrode surfaces and across interfaces (Fig. 1.11). The mechanistic understanding of these processes is still lacking, mainly due to the difficulty of characterizing these processes under in situ conditions. Electrostatic force microscopy (EFM) can be used in combination with AFM to map the carbon deposition on the nanoscale (<25 nm resolution). While AFM is capable of distinguishing the morphological variation after hydrocarbon treatment, EFM simultaneously gauges the surface potential through which the surface phases are identified. If for instance nickel and carbon are the two species present on the sample, each have a different surface potential, and so EFM is capable of separating them. Nanoscale species mapping based on EFM is very promising to study the impact of nano-sized catalysts applied onto the materials. By comparing the EFM images that show the local carbon deposition behavior for different catalysts, their resistance capabilities can be better evaluated.

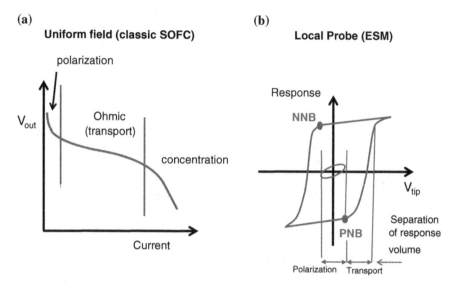

Fig. 1.11 a Schematics of current-voltage response of a classical fuel cell. **b** Equivalent nanoscale behavior illustrating minor and major ESM loop. Below the critical potential for activation of gas-solid reaction, the loop is closed. Above this potential, the loop opens. The positive and negative nucleation biases (PNB and NNB) provide local analogs for cathodic and anodic polarization

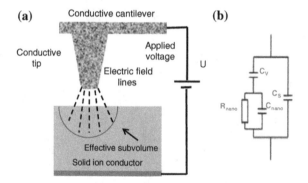

Fig. 1.12 a Schematic illustration of the experimental setup for electrostatic force spectroscopy on solid electrolytes. **b** Equivalent circuit for modeling the overall capacitance of the system

Electrical scanning force methods are characterized by an additional electrical potential difference between tip and substrate, and Electrostatic Force Microscopy (EFM) is a type of dynamical non-contact AFM where the electrostatic force is probed. Figure 1.12 shows the principle setup of an EFM. To good approximation the electric field around the tip decays radially.

Chapter 2
Emergent Local Probe Techniques

Abstract New local probe techniques continue to be developed for studying several particular interactions at the nanoscale and also as ideal tools for nano-patterning at this scale. Their applications in electrochemistry and friction phenomena have a special relevance for improving energy efficiency and energy storage devices. Friction represents a drawback in increasing energy efficiency performances of micro-electromechanical devices. The SPM techniques are also the ideal tools to study friction at the nanoscale and understanding the force that arises from the zero-point field may enable its control and so improving the non-contact motion of molecular motors and nanomachines. For sub-micron distance between two bodies, the Casimir force far exceeds the gravitational force. Thus, precision measurements of the Casimir force can be used for checking unification theories of the existence of extra dimensions along with a host of new particles that lead to deviations from Newtonian gravity in the 100 nm distance range.

Keywords Nanophysics · Nanoprobes · Nanotribology

2.1 In Situ Local Probes

The performance of electrochemical energy storage systems is limited by the fundamental behavior of the materials used in the cells, including electrode active materials and supporting components. Although often perceived as a simple electrochemical device, batteries are inherently complex dynamic systems. Their successful operation relies on a series of interrelated mechanisms, some involving instability of the components induced by charge/discharge cycles and formation/reaction of metastable phases. The ability to achieve long-term stability requires elucidation of the underlying physical and chemical processes. Thus, it is essential to use in situ non-invasive characterization tools and methodologies to study these processes in a wide variety of temporal and spatial resolutions.

© The Author(s) 2015
R.F.M. Lobo, *Nanophysics for Energy Efficiency*,
SpringerBriefs in Energy, DOI 10.1007/978-3-319-17007-7_2

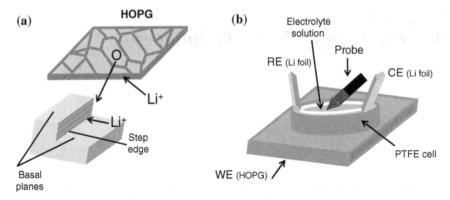

Fig. 2.1 Schematic diagrams of **a** highly oriented pyrolytic graphite and **b** electrochemical cell for in situ SPM observation

   In situ scanning probe microscopy for studies of electrode/electrolyte interface with high spatial resolution, has been developed and the decomposition of electrolyte to form the Solid Electrolyte Interphase (SEI) on Highly Oriented Pyrolytic Graphite (HOPG) has been discussed and images are used to show how the morphology of the SEI layer formation is correlated with the corresponding electrochemical process. Since HOPG has an atomically flat surface, it is often used as a model electrode for graphite negative electrode. A typical cell for in situ AFM measurements is simple, and can be constructed with a commercially available highly purified electrolyte solution and lithium foil as counter and reference electrodes (Fig. 2.1).

   Charge/discharge properties, cycleability, durability of Lithium-ion batteries (LIBs) are significantly affected by the kind of electrolyte systems. One of the reasons is attributable to the passivating surface film (SEI) that is formed on the graphite negative electrode. This film occurs by decomposition of the electrolyte solution upon the initial charging of LIBs. The SEI prevents further solvent decomposition, and enables lithium ions to be intercalated within the graphite negative electrode. Most of the commercially available LIBs employ solvent systems based on ethylene carbonate as a primary solvent, because it gives a stable and superior SEI. The SEI formation on graphite negative electrodes has been intensively investigated using in situ SPM methods.

   In LIBs it is common to use porous composite electrodes made of an active material powder, a carbon conductor, and a binder, coated on a metal current-collector (Cu or Al). However, in most cases, they cannot be used in SPM observation because the surface roughness is too large to detect small changes in morphology during charging and discharging. Then model electrodes with smooth surfaces (e.g., thin film electrodes prepared by sputtering, vacuum deposition or electrodeposition) can be used for in situ SPM measurements. Freshly cleaved surface of HOPG is a model of graphite negative electrode for investigating SEI formation. It can be shown that STM images of HOPG basal plane obtained at

2.1 V versus Li+/Li (where no reaction takes place), show usually steps of 3 nm in height (which means about 9 layers of graphene sheets); after the potential was kept at 1.1 V, part of the basal plane surface in the vicinity of the step edge usually raise by about 1 nm. An electronically insulating layer was not formed on the basal plane surface at 1.1 V, because the images are obtained by STM, which needs conductive surface for observation. The observed height of the hills (1 nm) is comparable to interlayer spacings in intercalation compounds of alkali metals with organic solvent molecules, such as tetrahydrofuran (THF) and dimethoxyethane (DME). It is thus likely that these hills were formed by influence of the solvent.

According to the proverb "*Seeing is believing*", in situ SPM enables the direct observation of the electrode/electrolyte interface and therefore is a powerful technique for understanding interfacial phenomena in LIBs. Understanding the interfacial phenomena is a key to the development of innovative active materials and solvent systems for LIBs. There has been a remarkable improvement in the SPM equipment for the past two decades, and even combinations with mass spectrometry have been set-up.

The local chemistry of several surface reactions is still unknown, since no technique exists for identifying atom types at individual sites on crystal surfaces. In some UHV surface science applications this is not a serious drawback, since surfaces of known structure are prepared with atomic cleanliness and changes due to the addition of one known additional species are studied. In Scanning Tunneling Atom Probe (STAP) the combination of STM with mass spectroscopy can be effective since it combines the high lateral resolution of STM topography with chemical identification [45]. In STAP a sharp STM tip is scanned over a surface and a short, positive voltage pulse is applied to the tip, causing some atom transfer from sample to tip. The sample is then removed, and a much larger voltage pulse is applied, causing field desorption of the atoms which have been transferred from the sample to the tip. The flight time of the atoms from tip to detector is measured as in the atom probe, giving their mass to charge ratio [45]. STAP has been applied in chemical identification of atoms captured on an STM tip during atom manipulation as in the case of direct evidence of tungsten silicide formation during atom manipulation was clearly shown [46].

The main problem with STAP in addition to its inherent complex and laborious experimental UHV procedures concerns the diffusion of atoms on the probing tip; since metal atoms diffuse rapidly away from the region of highest curvature one needs to cool the probing tip.

Also spectroelectrochemical techniques used for the investigation of battery components, benefited from the advances of Raman technology like the implementation of confocal Raman microscopy in the battery research, which opened the way to new and more sophisticated applications. The discovery of new Raman techniques such as Surface-enhanced Raman scattering (SERS), Tip-enhanced Raman spectroscopy (TERS), spatially offset Raman spectroscopy as well as the integration of Raman spectrometers into non-optical microscopes, for example AFM and SEM, allowed to perform two or more analytical techniques on the same sample region, with a very high resolution. All these progresses led to new insights into battery materials and components such as electrodes and electrolytes, and helped to understand the electrode/electrolyte interface phenomena like battery aging and the dynamic nature of the solid electrolyte interfaces in lithium batteries.

Despite this book is focused on SPM one wants to mention briefly that other local probe techniques for electrochemical energy storage systems are also relevant such as the synchrotron X-ray diffraction (XRD) and in situ XPS (X-ray photoelectron spectroscopy).

The lithium-ion battery is one archetype of system whose performance and durability have significantly benefitted from of synchrotron radiation studies. Advanced X-ray techniques with synchrotron radiation have been demonstrated on lithium battery materials where $LiMn_2O_4$ spinel and related ceramic materials are used as cathodes, and structural changes during (dis-)charging are monitored in situ with XRD. Upon charging with lithium, manganese oxide undergoes a phase transformation which is considered a major origin of electrode failure.

Synchrotron-based X-ray diffraction and advanced electron microscopy were also used to characterize the unique structural features of mixed-valent manganese oxides. The unique mixed-valent (2+, 3+, and 4+) manganese oxides with porous nano-architecture are well suited for fast mass and charge transfer associated with energy storage processes.

Nanostructured mixed-valent manganese oxides were successfully deposited onto porous carbon fiber paper (CFP) using a simple precipitation method in an aqueous solution at low temperature (75 °C) followed by a controlled annealing process to obtain the desired composition and microstructure [48]. Gas adsorption/desorption analysis showed that the microstructure of MnO_x was dramatically rearranged during the annealing process, producing porous, nano-structured MnO_x.

Regarding the XPS technique it has been successfully applied in combination with electrochemistry for in situ investigations, in particular for electrochemical double layer capacitors (EDLCs). The boom of mobile and portable electronic devices enhance the need of new highly performing energy storage devices and this triggers the research of high-energy/high-power storage devices. Electrochemical double layer capacitors (EDLC) are high power devices bridging the gap between conventional capacitors and batteries. The energy and power density of EDLC is dependent on the square of the applicable cell voltage, enhancing the research for more stable electrolytes for these devices. Nowadays, the cell voltage of commercially available EDLC is limited to 2.7 V, mainly due to faradaic reactions of the organic solvents of the electrolytes at the High Surface Area Carbon (HSAC) electrodes. Ionic liquids are a new type of solvent free electrolytes and they are known to have an increased electrochemical stability window. Reliable electrochemical studies require a stable reference electrode and activated carbon has been shown to serve as stable reference electrode for the use with different ionic liquids. A combination with Density Functional Theory (DFT) calculations and X-ray photoelectron spectroscopy (XPS) has revealed, that the valence band structure of ionic liquids can be correlated to the electrochemical stability. Ionic liquids have negligible vapour pressure and are compatible to UHV analytical tools, such as

XPS. Therefore, an in situ combination of a surface sensitive technique like XPS and ionic liquids electrolytes is a promising tool to study electrochemical processes in situ.

For that purpose, in situ electrochemical XPS cell have been developed. D'Agostino and Hansen [49, 50] and Kolb [51] established the so-called electrode emersion technique, which can be described as a quasi in situ analysis of the electrode/electrolyte interface. In general, an electrochemical reaction chamber is attached to the distribution chamber of the XPS system, allowing for electrochemical measurements under inert atmosphere and subsequent sample transfer of the working electrode to the XPS analysis chamber without contact to ambient air (Fig. 2.2). Using this technique several interesting electrochemical reactions in aqueous electrolytes, such as O_2 evolution [53], oxide formation [54], or semiconductor etching [55] could be successfully analysed. It was even found that by proper emersion of the electrode, the Electrochemical Double Layer (EDL) stays intact and the so-called electrochemical shift was observed for species present in the EDL by XPS analysis [56]. This shift, first discussed by D'Agostino and Hansen [49, 50] reflects the influence of an electrochemical potential on the electron binding energies for species present in the electrochemical double layer. Species present at the outer Helmholtz layer will have their energy levels fixed to the solution side, whereas species in the inner Helmholtz layer experience a change in binding energy according to a change in the work function of the metal working electrode [56].

The first report about an in situ XPS cell was published by the group of Licence [57] in 2009. A two electrode arrangement, using the beaker for the ionic liquids as counter (and reference) electrode and a metal wire working electrode is introduced into the XPS analysis chamber (Fig. 2.3).

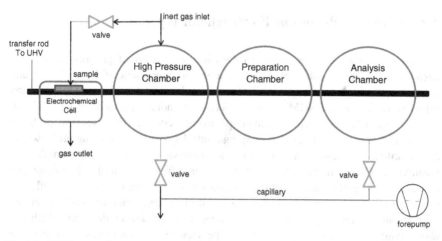

Fig. 2.2 UHV system incorporating a quasi in situ chamber with electrochemical cell and sample to be probed by surface analysis techniques including XPS

Fig. 2.3 Sketch of a typical in situ XPS electrochemical cell

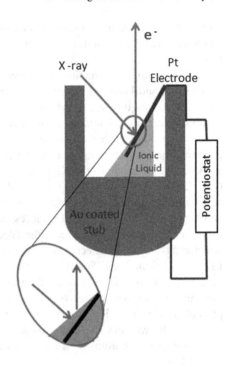

Using that setup, the electrochemical reduction of Fe^{3+} to Fe^{2+} in a mixture of [EMIM][EtOSO$_3$] and [BMIM][FeCl$_4$] was monitored in situ. High resolution scans in the Fe region were successively taken by monitoring the reduction of Fe^{3+} species to Fe^{2+} species [57].

2.2 Scanning Probes in Electrochemistry

Scanning Microwave Microscopy (SMM) has seen a lot of progress in the last decade. A variety of different designs for the measurement of both electric and magnetic properties for a wide range of materials have been reported. The operation principle is similar to SNOM but uses microwave radiation instead of visible light.

The near-field microwave microscope mainly consists of a sensing probe, a microwave detection unit, and an imaging control system. The sensing probe works as an antenna structure that transmits a microwave signal to the sample and meanwhile collects the reflected microwave signal from the sample; the microwave detection unit generates the source microwave signal and monitors the reflected signal simultaneously while the imaging control system scans the probe over the sample surface and regulates the tip-sample distance during the scan. With the above configuration, the microscope can be viewed as a microwave transmission line terminated at the end of the sensing probe, with a terminal impedance Z.

As a result, SMM can be used to measure the capacitance at the near surface region, and to perform free carrier concentration profiling on a semiconductor device. It becomes possible to probe only a small subsurface region of a normal metal sample using near-field micro-wave microscopy.

Scanning Ion Conductance Microscopy (SICM) is a SPM technique which is well-suited to topographic and chemical mapping of physical interfaces. In SICM, a potential difference applied between an electrode inside an electrolyte-filled nanopipette and a second electrode outside the nanopipette results in a steady-state ion current. This ion current flowing through the pipette is strongly influenced by the relative position between the SICM probe and a sample of interest, providing a feedback signal to precisely control the position of the pipette. These position-dependent changes in conductivity enable SICM to measure both nanoscale features and physical properties of the sample under study.

On its turn, Scanning Electrochemical Microscopy (SECM) is a type of scanning probe microscopy that uses faradaic current (i.e. current produced by electron transfer in a redox reaction) to obtain information about a system under study.

Scanning Ion Conductance Microscopy (SICM) combined with Scanning Electrochemical Microscopy (SECM) certainly represent key tools for the electrochemical characterization of interfaces used in energy applications. Although they are conceptually similar in some respects, and share some hardware and scanning software, the SECM and SICM complement each other in each specific situation and reach current sensitivities of sub-pA levels. For example, in the particular case of in situ characterization of lithium ion batteries, the SICM is better-suited than the SECM, as the latter requires that the electrolyte be decomposed for ion flow, interfering with standard battery operation.

In its fundamental operating mode, the SICM uses two electrodes immersed in a conductive solution (usually 0.1 M NaCl), measuring the current between them: one of the electrodes is inside a pipette, while the other is relatively far from any object. When the pipette is far away from the sample to be investigated (a sample that is also located within the solution) then a current maximum is measured. Next, the pipette is automatically lowered towards the sample: at a certain tip-sample probe distance the current begins to decrease due to a physical restriction of the pipette tip aperture, resulting from increasing proximity of the sample being investigated.

The SICM uses micro and nanopipettes, whose tip diameter can approach the nanoscale. In conventional SICM, the pipette electrode, counter electrode and sample are all in the same solution. Actuation in Z direction moves the pipette towards the sample, while a patch-clamp amplifier (I) and control software and data acquisition (II) operate the scanning protocol. There are also modified variants where the bath solution is eliminated, and the substrate acts as the working electrode.

Many of the fundamentals featured in SICM instruments are shared not only by other SPM techniques but also by simple electrochemical workstations: a potentiostat, amplifier, the working, counter and reference electrodes, the electrochemical cell, electrolyte solution and software for procedures such as a scanned voltammogram. The associated data acquisition hardware and wiring is also conceptually similar, albeit sophisticated components are required for high-speed real-time SPM operation. The precise positioning of the SICM nanopipetteis assured by movement through piezoactuators. This allows the tip to be scanned accurately and rapidly, enabling spatially controlled electrochemical analysis. The isolation from external disturbances such as electrical noise and physical vibrations can be achieved through a Faraday cage and air table, respectively.

The probes used by SICM instruments can be glass capillaries pipettes whose tip apertures are in the order of microns ("micropipettes") or nanometers ("nanopipettes"). The resolution of SICM instruments is in large part determined by pipette tip aperture.

The development of ultra-microelectrodes and nanopipettes is currently progressing towards custom functionalization. In fact, an important advantage of the SICM pipette fabrication workflow is that capillary glass can be functionalized, extending the range of applications. For example, for operation where droplet spreading must be confined, the outer glass surfaces of the pipettes can be silanized, useful for operation at air/liquid and liquid/liquid interfaces. Other more complex pipette modification examples include an aldehyde-terminated layer for enabling imine bonding and ultimately resulting in a highly cationic surface charge, useful for sensing applications.

One relevant advantage of the SICM is its ability to characterize with high resolution samples that are soft and that display a large and stimulus-responsive variance in height, such as live cells, which conventional SPM techniques have difficulty with. Since the separation between SICM tip and soft substrate is only a few microns, the measured ion exchange is representative of a relatively precise area.

Recently there has been progress towards the combination of SECM with SICM into a unique technique through the convergence of the SECM ultramicroelectrode and SICM nanopipette into a single scanned microscopy probe. The resulting combined electrode generated both electrochemical and ionic current signals, and was capable of simultaneous SECM and SICM curves; however it suffered from the disadvantage of increasing diameter due to multiple layer deposition. Specifically, the final SECM/SICM probe diameter is in the order of 500 nm and limited to submicron resolution rather than true nanoscale topographical and electrochemical imaging. Another approach for combining SECM/SICM probes is based on double-barreled nanopipettes: one barrel was left hollow as in conventional SICM, filled with an aqueous electrolyte and electrode; the other is filled with butane gas that is subsequently converted to carbon through pyrolytic decomposition. The resulting combined SECM/SICM probes is called "double carbon nanoprobes" (DBCNPs), and has the advantage of rapid manufacture and reduced diameter (under 200 nm). The demonstration of successful simultaneous topographical and electrochemical imaging with high resolution was done on various substrates (i.e., polyethylene terephthalate, platinum and living neurons).

An AC feedback signal allows for topography to be determined, and a DC current can be used to infer the spatial variation in local ionic current resulting from material inhomogeneities. One such result is shown in Fig. 2.4, where the

Fig. 2.4 Experimental setup of an SICM configuration

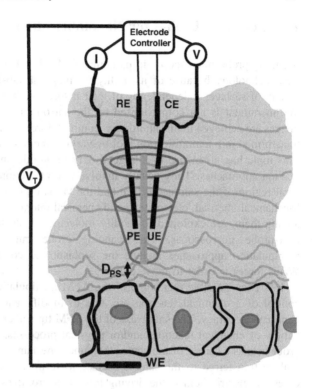

electrochemical characterization of silicon nanoparticles and Polyvinylidene difluoride (PVDF) material on a copper foil is shown.

In addition, to study the effect of electrode material thickness on electrochemical response, a topic of interest in battery technology is the formation of a solid electrolyte interphase, which reduces local ionic current. The SICM investigation has found that drops in current were in fact not necessarily related to changes in topography, but can be attributed to the formation of a solid electrolyte interphase.

In polymer membranes and catalysts for future energy applications such as fuel cells, with their selective proton transport membranes, the precise and localized measurement of flux through pores is of key importance. The SICM is inherently well-suited to these types of measurements due to its design. Indeed, the SICM could localize individual pores, including the mapping of single active ion channels (i.e., in cell plasma membranes).

As the scanned probe microscopy techniques of SECM and SICM develop, the need for correlation with additional characterization techniques will be required, namely imaging via SEM or using Raman Spectroscopy (RS) and Fourier Transform Infrared Spectroscopy (FTIR).

2.3 Scanning Probes for Nanofabrication

Scanning probe microscopy, in particular AFM, is a key tool in Nanoscience and Nanotechnology because of its ability to image at sub-10 nm resolution a wide variety of surfaces independently of their nature. Furthermore, the performance of the instrument is not compromised by the surrounding medium. High resolution images are achieved in air, liquid or vacuum. A force microscope can also be easily transformed into a modification tool by varying the relevant tip-surface interaction. This topic has given rise to a large variety of atomic and nanometer scale modification approaches. They usually involve the interaction of a sharp probe with a local region of the sample surface and the variation of one or several parameters. Mechanical, thermal, electrostatic and chemical interactions, or some combinations among them, are exploited to modify molecules, nanostructures or surfaces with near-probe microscopes. The scanning probe microscopy modification and manipulation approaches include the sophisticated control of attractive van der Waals forces in order to move atoms.

It is even possible that some of the AFM manipulation methods based on the control of a chemical reaction are able to modify and/or pattern surfaces with nanoscale accuracy. The small size of the AFM tip's apex is often used to confine a variety of chemical reactions and/or physical processes. In some cases, the AFM probe acts as a carrier of molecules that, upon mechanical contact with the surface, will be transferred to the sample surface to form new chemical bonds. In other cases, an electric field is the driving force that promotes the field-induced evaporation of atoms from the tip. Other processes are mediated by the presence of a liquid meniscus which provides both the chemical species and the spatial confinement for a chemical reaction to occur. Finally, in some cases a hot tip facilitates the breaking of chemical bonds on certain polymer surfaces.

The relationship between AFM and chemistry was initially motivated by nanopattering applications, and recently new routes expand this relationship to other contexts. In field-induced chemistry, unusual chemical reactions have been observed in the presence of very high electric fields (10–50 V/nm) with field ion microscopes. Such fields are of the same order of those inside atoms and molecules. Consequently, they are strong enough to induce the rearrangement of molecular orbitals leading to new chemical reactions.

A force microscope interface offers a precise control and manipulation of high electric fields by changing the conductive tip-surface separation and/or the voltage (Fig. 2.5). In contrast to field-ion microscopy that operates in ultra-high vacuum, the AFM experiments can be performed in ambient or liquid environments, which increases the number of available chemical species. Remarkably, high electric fields can be achieved in AFM by applying moderate voltages. The electric field between two flat conducting surfaces separated by a distance D can be estimated as $F = V/D$. The presence of a sharp probe introduces a correction factor to the above expression. For a hyperboloidal tip, the field at the tip's apex is

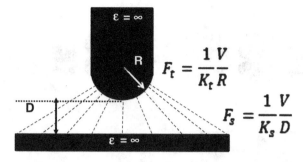

Fig. 2.5 Electric field between an AFM conductive tip of radius R and a flat surface separated by a distance D. A radial distribution is assumed

$$F_t = \frac{1}{K_t}\frac{V}{R} \qquad (2.1)$$

while at the flat surface

$$F_s = \frac{1}{K_s}\frac{V}{D} \qquad (2.2)$$

where K_t and K_s are geometrical factors.

If $R \ll D$, it is deduced that

$$K_t = K_s = \frac{1}{2}ln\frac{4D}{R} \qquad (2.3)$$

where V is the applied voltage and R is the probe radius; $K_t \approx 2$ and $K_s \approx 3$ are some values representative of many AFM experiments.

In a force microscope interface the field can be easily determined and varied by either modifying the tip-surface distance or the voltage. Since the tip-surface separation is in the nanometer or sub-nanometer range, very high fields (1–30 V/nm) can be generated in the presence of tens of volts. In the presence of a gas, liquid, or solid material, those fields could be used to promote the evaporation of atoms from the tip, the breaking of chemical bonds in the molecules of the surrounding environment or the formation of new products. Thus, the success of the relationship between field-induced chemistry and nanofabrication relies on the control and manipulation of electric fields at the nanoscale. In particular, electric field-induced dissociation of carbon dioxide can subsequently be used for generation of a carbonaceous pattern with an AFM conductive tip (Fig. 2.6).

The activation of carbon dioxide by electric fields could be then up-scaled by using stamps patterned with billions of nanoscale asperities.

The transformation of CO_2 in the presence of an electric field can be separated into three major steps: the first one is the capture of the molecules into the gap region between the two electrodes; the second step is the activation of the molecules

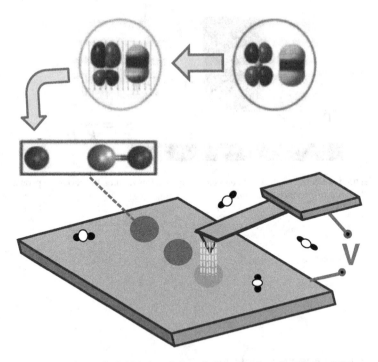

Fig. 2.6 Schema of electric field-induced dissociation of carbon dioxide by SPM

and the third corresponds to the fabrication of a carbonaceous compound. The first step is a field-induced diffusion process. The electric field polarizes the carbon dioxide and the resulting dipole moment interacts with the field. The potential energy of a CO_2 molecule under the influence of the field F can be approximated by

$$U = U_0 - \frac{1}{2}\alpha F^2 \tag{2.4}$$

where $\alpha = 2.93 \times 10^{-40}$ C^2 m^2 J^{-1} is the static polarizability of CO_2 and U_0 is its free energy in the absence of the field. Thus, in a non-uniform field, the molecule will experience a polarization force towards the conductive surfaces. The presence of high electric fields (0.1–30 V/nm) in the interface will trap the CO_2 between the conductive surfaces, and thus increase the gas density. One can estimate the change of pressure, p, between the electrodes as:

$$p = p_0 \exp\left[\frac{\Delta U}{k_B T}\right] \tag{2.5}$$

where p_0 is the pressure at zero field. For a field of 10 V/nm, one gets $\Delta U = 92$ meV which in turns gives $p \approx 36$ p_0 (T = 298 K). The increase of pressure implies a higher collision rate between the gas molecules.

Quantum chemical calculations have shown that the field modifies the potential energy surface of dissociation into C0 + 0 and shifts the energy of the molecular orbitals which leads to carbon dioxide splitting. The field shifts the energy of the molecular orbitals and modifies the dipole moment. As the field increases, the molecular dipole moment builds up and different local charges form on the atoms. The carbon atom becomes more positive, while the charge difference between the two oxygen atoms increases. Also the two bond lengths become asymmetric (0.122 nm vs. 0.113 nm at 30 V/nm). The field also greatly affects the Lowest Unoccupied Molecular Orbital (LUMO), which becomes nearly degenerate with the Highest Occupied Molecular Orbital (HOMO) at 40 V/nm. At high fields, the HOMO becomes located outside the molecule. Closing the HOMO-LUMO gap promotes the detachment of the oxygen atom and the formation of two fragments. Calculations, at fields higher than 40 V/nm, show that the molecule spontaneously breaks down into C0+0. Concomitantly, the field affects a sigma molecular orbital, which becomes located on a CO fragment until it closely resembles the HOMO of carbon monoxide. Under these conditions, the products are charged and can easily react to form a solid, carbonaceous material, as it was confirmed by X-ray photoelectron spectroscopy (XPS). Scanning probe microscopes have also been used to investigate other types of carbon dioxide processes such as the photocatalytic dissociation of CO_2 on titanium dioxide surfaces.

The high electric field in the proximity of the tip's apex has also been used to induce the evaporation and/or ionization of a variety of materials. The experiments involving ionization have two steps: First, the molecules are ionized and ejected from a tip that is biased negatively (-6 to -10 V); second, the ejected molecules are electrostatically attached to an attractive surface. The patterning can be switched on and off by applying a negative bias or by turning it back to zero, respectively. On the other hand, the dimensions of the patterns can be controlled by modulating the tip bias and the scanning speed. This method has been applied to deposit different kinds of organic materials, such as fullerenes, polymers, and inorganic materials such as gold on different substrates: HOPG (highly oriented pyrolytic graphite), graphite, ITO (indium tin oxide), or gold.

AFM Oxidation, also known as tip-based oxidation or local oxidation nanolithography, provides a remarkable example of the relationship between electric fields, chemistry and probe microscopy. AFM oxidation is based on the spatial confinement of an anodic oxidation between the tip and the sample surface. In air, the oxidation process is mediated by the formation of a nanoscale water bridge (Fig. 2.7).

The bridge acts as a nanoscale electrochemical cell. The tip is biased negatively (cathode) with respect to the sample surface (anode). AFM oxidation can be either performed with the tip in contact with the sample surface or in a non-contact mode. In the latter case, the formation of a water bridge requires the application of a voltage. The non-contact AFM oxides are smaller and show higher aspect ratios under similar experimental conditions. Another advantage of the dynamic mode is that the tip suffers less wear and its lifetime is increased. This is an important point to address pattern reproducibility. Tip-based oxidation started with the experiments

Fig. 2.7 Schema of SPM oxidation highlighting the role of the water meniscus in the redox reactions at the tip-sample interface

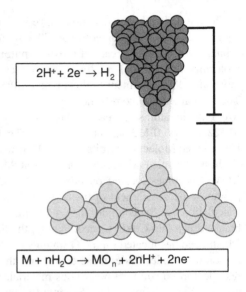

performed with a scanning tunneling microscope (STM). However, the STM is rarely used today for local oxidation purposes because of the difficulties associated with its operation on poorly conductive surfaces. In AFM oxidation, the electric field plays three roles: First, it induces the formation of a water bridge; second, it generates the oxy-anions needed for the oxidation by decomposing water molecules; third, it drives the oxy-anions to the sample interface and facilitates the oxidation process. On silicon surfaces, the local oxidation process is able to generate ultra-small silicon oxide nanostructures with a minimum lateral size of about 12 nm and a height that ranges from 1 nm to tens of nm, depending on the oxidation conditions. The height and the width of the oxide depend linearly on the voltage. The height h also shows a power law dependence with the pulse time t of the type $h = bt^{\gamma}$, where γ is in the 0.1–0.3 range for Si(100) surfaces. Voltage pulse amplitudes and durations are, respectively, in the 10–30 V and 0.005–1 s ranges. The role of the water meniscus is two-fold. It acts as a nanoscale electrochemical cell that provides the oxy-anions thanks to which the reaction takes place. In addition, it confines the reaction laterally, i.e., the size of the meniscus determines the resolution of the features obtained by this technique. During the redox process, the tip acts as the cathode, generating hydrogen:

$$2H^+(aq) + 2e^- \rightarrow H_2 \tag{2.6}$$

The anodic reaction occurs at the surface of the metallic or semiconductor substrate as follows:

$$M + nH_2O \rightarrow MOn + 2nH^+ + 2ne^- \tag{2.7}$$

The local oxidation process is accompanied by an extremely small faradaic current. The current reflects the flux of O^- and OH^- ions between the tip and the surface and it can be monitored in situ during the oxidation process. The current flow usually shows a quick increase upon liquid bridge formation (and tip approach). Then it begins to decrease while the oxide is being built. In the latter stage the current values are in the pA and sub-pA range. One should also mention that growth of nanostructures by AFM oxidation is space-charge limited due to the liberation of ions during oxidation.

The existence of a negative charge within the oxides can be exploited to use the local oxides as templates for the organization of positively charged molecules (such as ferritin). The main parameters that control the local oxidation lithography process are the applied voltage (from a few volts to 30 V), the relative humidity, the duration of the process (10 p.s–10 s), the tip-sample distance (2–5 nm) and the scan speed (0.5 μm/s–1 mm/s). The presence of defects that are created during the growth of the oxide will also affect the final size of the patterns. AFM oxidation has been used to create a variety of nanoscale patterns and nanomechanical or nano-lectronic devices.

Figure 2.8 shows the steps to build some molecular architectures by combining AFM oxidation and self-assembling methods. First, the silicon surface is functionalized with an octadecyltrichlorosilane (OTS) monolayer. This replaces the OH^- groups that dominate the native silicon surfaces in air with a methyl-terminated neutral surface. Then, a region of the functionalized surface is locally oxidized. The oxidation process removes the OTS monolayer. The sample is immersed in an aminopropyltriethoxysilane (APTES) solution until an APTES monolayer is deposited on the patterned region. The last step involves the deposition of the proteins. An example of the above process is the deposition of single molecules of ferritin on silicon surfaces with an accuracy similar to the size of the molecules (10 nm). The AFM images of the nano-stripes taken before and after the deposition of ferritin molecules are shown in Fig. 2.9. There is a total absence of ferritin molecules outside the patterned stripes. This result underlines the strong selectivity of the patterning process. Sophisticated nano-electronic devices such as silicon nanowire (SiNW) transistors and circuits have been fabricated by AFM oxidation nano-lithography. In this application, AFM oxidation generates a narrow

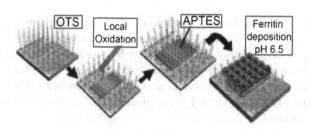

Fig. 2.8 Scheme of the main steps followed to obtain a ferritin pattern on a silicon surface by combining bottom-up electrostatic interactions and AFM oxidation. (reprinted with permission [58])

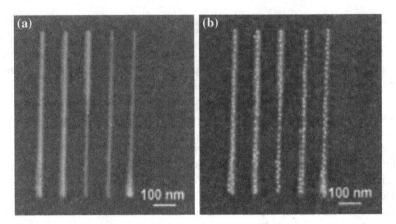

Fig. 2.9 a Silicon oxide nanostripes fabricated by AFM oxidation. **b** Selective deposition of ferritin molecules on the AFM oxidation pattern. (reprinted with permission from [58])

oxide mask on top of the active layer of a silicon-on-insulator substrate. The unmasked silicon layer is then removed by using wet or dry etching. The local oxide protects the underneath silicon from the etching. This leaves a single-crystalline silicon nanowire (SiNW) with a top width that matches the width of the oxide mask. SiNWs with a channel width of 4 nm have thus been fabricated. These types of nano-electronic transistors are promising in the development of very sensitive bio-sensors.

Silicon and titanium are the most used materials to perform local oxidation experiments; nonetheless, local oxidation has been extended to other interesting materials such as SiC or graphene. Selective oxidation of Self-assembled Monolayers (SAMs) to build molecular architectures has also been accomplished by AFM oxidation, in a combination of bottom-up and top-down nanofabrication processes, and has the potential to generate complex interfaces.

2.4 Advanced Scanning Probes and Nanotribology

We have already discussed the applications of SPM in imaging surface topography, measurement of local properties on sample surface and nanolithography in processing of nanodevices. Now we will proceed by discussing other relevant applications of scanning local probes as nanodevices, and in nano-locomotion, and address that the ever present friction phenomena at the nanoscale lead us ultimately to the quantum vacuum realm and to the practical limitations of future electromechanical nanodevices.

When the surface of a microprobe is functionalized in such a way that a chemically active and a chemically inactive surface are morphologically alternating, then chemical or physical processes on the active cantilever surface can be detected using the temporal evolvement of the probe's response. Cantilevers can be used as a nanomechanical sensor device for detecting chemical interactions between binding partners on the cantilever surface and in its environment. Such interactions might be produced by electrostatic or intermolecular forces. At the interface between an active cantilever surface and the surrounding medium, the formation of induced stress, the production of heat or a change in mass can be detected. In general, detection modes can be grouped into three strands: static mode, dynamic mode and heat mode as illustrated in [60]. In the static mode, the bending of the cantilever beam due to external influences and chemical/physical reactions on one of the cantilever's surfaces is investigated. The asymmetric coating with a reactive layer on the surface of the cantilever favors preferential adsorption of molecules on this surface. In most cases, the intermolecular forces in the adsorbed molecule layer produce a compressive stress, i.e. the cantilever bends. If the reactive coating is polymeric and adsorbing molecules can diffuse, the reactive coating will swell and the cantilever beam will also bend. Similarly, if the cantilever beam emerges into a chemical or biochemical solution, the asymmetric interaction between the cantilever beam and the surrounding environment results in bending of the cantilever beam.

There are many new concepts and devices that have been explored. In dynamic mode, the cantilever is driven at its resonance frequency. If the mass of the oscillating cantilever changes owing to additional mass deposited on the cantilever, or if mass is removed from the cantilever, its resonance frequency varies. Using electronics designed to track the resonance frequency of the oscillating cantilever, the mass changes of the cantilever are derived from shifts of resonance frequency. The cantilever can be regarded as a tiny microbalance, capable of measuring mass changes of less than 1 pg. In dynamic mode, active coatings should apply on both surfaces of the cantilever to increase the active surface where the mass change takes places. Dynamic mode works better in gas than in liquid, which complicates the exact determination of the resonance frequency of the cantilever. In heat mode the cantilever is coated asymmetrically, one surface with a layer having a different thermal expansion coefficient than that of the cantilever itself. When such a cantilever is subjected to a temperature variation, it will bend and deflections corresponding to temperature changes in the micro-Kelvin range can be easily measured. If the coating is catalytically active (e.g. a platinum layer facilitates the reaction of hydrogen and oxygen to form water) and heat is generated on the active surface, then it will result in bending of the cantilever. Such a method can also be used in the study of phase transitions and measurement of thermal properties of a very small amount of materials. Although the above discussion has been limited on the single cantilever nanosensors, the same principle is also applicable to multiple cantilever nanosensors.

Nanotribology studies friction phenomenon at the nanometer scale, where atomic forces absolutely determine the final behavior of the system. As a common sense, gears, bearings, and liquid lubricants can reduce friction in the macroscopic world, but the origins of friction in small devices such as micro- or nano-electromechanical systems (MEMS/NEMS) require other solutions. These systems exhibit enormous surface-volume ratio that leads to severe friction and wear. Therefore, this dramatically reduce their applicability and lifetime.

Traditional liquid lubricants become too viscous when confined in layers of molecular thickness, and so other solutions for reducing friction on the nanoscale (such as superlubricity and thermolubricity), have been highlighted.

The major experimental tool for modern use in Nanotribology research is the Atomic Force Microscope (AFM), which has several variations. The dynamics of the interactions of two surfaces during relative motion, ranging from the atomic level to microscale, need to be understood in order to develop a fundamental understanding of friction. In case of relative motion in contact, it means from the microscopic point of view that friction originates at multiple contacts, and so investigating single-asperity contacts assumes high relevance (Fig. 2.10).

AFM is capable of investigating conductive and non-conductive surfaces at the atomic level and AFM experiments have revealed that the relationship between friction and surface roughness is not always obvious or straightforward. Friction also occurs when there is no contact between the two bodies in relative motion because interatomic forces are always present at the nano level, even if the two surfaces are neutral from the electrical point of view.

An AFM tip can play the role of a single-asperity contact with a solid or lubricated surface (Fig. 2.10) and the AFM microscope can measure ultra-small forces (<1 nN) in the scan direction, between the tip mounted on a flexible canti-lever beam and the sample surface. AFM microscopes can be used to study tri-bological phenomena and various types of friction at the nanoscale, in the understanding that friction is a force resisting the relative motion of two compo-nents of a system. In fact, the AFM provides a very sensitive method for measuring ultra-small forces between a probe tip and a sample surface.

Subsequent modifications of the AFM led to the development of Friction Force Microscopy (FFM), designed for atomic- and microscale studies of friction. Such instrument sensor is also able to measure the lateral force (friction force) and com-mercial AFM/FFM is routinely used for simultaneous measurements of surface roughness and friction force (Fig. 2.11). Normal and frictional forces being applied at the tip-sample interface are measured making use of a laser beam deflection tech-nique. The laser beam reflected from the vertex of the cantilever is directed through a

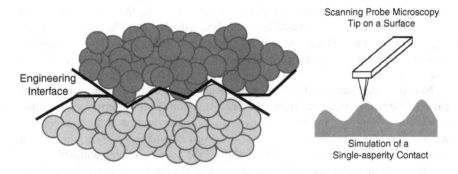

Fig. 2.10 Two surfaces in contact (*left*) and AFM tip in contact (*right*)

(a)   **(b)**

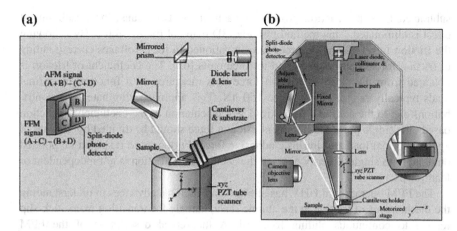

Fig. 2.11 Schematics of commercial atomic force microscope/friction force microscope AFM/FFM **a** of a small-sample, and **b** of a large-sample. (reprinted with permission [61])

mirror onto a quad-photodetector. The differential signal from the top and bottom photodiodes provides the AFM signal, which is a sensitive measure of the cantilever vertical deflection. Topographic features of the sample cause the tip to deflect in the vertical direction as the sample is scanned under the tip. This tip deflection will change the direction of the reflected laser beam, changing the intensity difference between the top and bottom sets of photodetectors (AFM signal).

In the AFM operating mode called "height mode" (for topographic imaging or for any other operation in which the applied normal force is to be kept constant), a feedback circuit is used to modulate the voltage applied to the piezo tube (PZT) scanner in order to adjust the height of the PZT, so that the cantilever vertical deflection (given by the intensity difference between the top and bottom detector) will remain constant during scanning. The PZT height variation is thus a direct measure of the sample surface roughness.

For measuring the friction force at the tip surface during sliding, left-hand and right-hand sets of quadrants of the photodetector are used. In the so-called friction mode, the sample is scanned back and forth in a direction orthogonal to the long axis of the cantilever beam. The friction force between the sample and the tip will produce a twisting of the cantilever, and as a result the laser beam will be reflected out of the plane defined by the incident beam and the beam reflected vertically from an untwisted cantilever. This produces an intensity difference of the laser beam received in the left-hand and right-hand sets of quadrants of the photodetector. The intensity difference between the two sets of detectors (the FFM signal) is directly related to the degree of twisting and hence to the magnitude of friction force.

One problem associated with this method is that any misalignment between the laser beam and the photodetector axis would introduce error in the measurement. However, by following the procedures developed by Bhushan et al. [62], in which the average FFM signal for the sample scanned in two opposite directions is subtracted from the friction profiles of each of the two scans, the misalignment effect is eliminated. This method provides 3D maps of friction force. By following the friction force calibration procedures mentioned in [62], voltages corresponding to friction forces can be converted to force units [63]. The coefficient of friction is obtained from the slope of friction force data measured as a function of normal loads typically ranging from 10 to 150 nN. This approach eliminates any contributions due to the adhesive forces [64]. For calculation of the coefficient of friction based on a single point measurement, friction force should be divided by the sum of applied normal load and intrinsic adhesive force. Furthermore it should be pointed out that, for a single-asperity contact, the coefficient of friction is not independent of load [62].

The FFM operating in UHV has allowed considerable advances in understanding the mechanisms of friction at the atomic scale, namely the transition from stick-slip regime to continuous sliding [65, 66]. A theoretical description of the FFM experiments, in terms of a model based on Brownian motion [65, 67] allows to obtain numerical simulations of the type shown in Fig. 2.12, where v is the velocity of the sample (with lattice constant of 0.5 nm) with respect to tip radius of curvature 10 nm, which is 3 μs^{-1}. F(t) is the instantaneous friction force versus the substrate position given by the product (vt). The motion of the tip along the axis xx, x(t) = F (t)/K (where K is the spring constant of the probe) has a *stick-slip* type characteristics in which the distance between "sticky" adjacent sites reproduce the lattice constant. Therefore it is simulated an atomic resolution, although the tip radius is an order of magnitude higher than the lattice constant. So, apparently, the *stick-slip* motion has its origin in the periodic potential of tip-substrate interaction, being the sliding an event of transition from one local minimum in the potential interaction

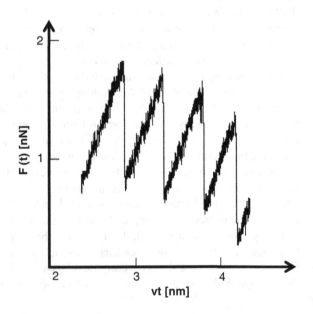

Fig. 2.12 Simulation of the instantaneous friction force F (t) versus the position of the substrate. The random fluctuations and not perfect periodicity of sliding events are a consequence of the role played by thermal fluctuations, that the model based on Brownian motion allows to reproduce [65, 67]

curve to another minimum in the neighborhood, caused by the elastic deformation forces. The periodicity of this potential corresponds to the lattice constant.

Friction is indeed a universal phenomenon observed during relative motion of surfaces, in all spatial scales (from macro to nano). Taking this in consideration and focusing in the rapidly emerging MEMS devices (micro-electrical-mechanical-systems), it is obvious that mapping the frictional force at the sub-micrometer scanning range is absolutely indispensable. Friction represents a major cause of loss of energy efficiency in systems consisting of moving parts. In these systems the friction is lower in the empty spaces between the moving parts, but even in these cases there is always a resistance to motion due to the omnipresence of inter-atomic forces. Therefore, the determination of these friction forces at the nano-scale using the AFM in non-contact mode is particularly relevant.

Contrarily to the AFM operating in contact regime (where a static deflection of the tip is recorded), in non-contact regime one needs to operate in dynamic mode, where the tip is kept in oscillation through a piezo-system connected to the cantilever, and the frequencies are measured by a phase-locked (PLL) electronic cycle. In this detection frequency operation the probe is excited at a frequency close to its resonance frequency, and the frequency change is measured as the interaction indicator parameter (Fig. 2.13). The tip-sample distance is kept in the interaction attractive regime (the reason for the term "non-contact": NC-AFM), and so this operation mode is completely non-intrusive contrarily to the tapping mode [61]. The NC-AFM requires the use of a cantilever with high quality factor Q.

One can consider a dynamic model of the forced and damped harmonic oscillator, in which a friction force proportional to the cantilever velocity is introduced in addition to the Hooke's law, and where γ is the damping coefficient (Amonton's law). The equation of motion for the position $z(t)$ of the probe is then given by:

$$m^* \frac{d^2z(t)}{dt^2} = -\gamma \frac{dz(t)}{dt} - kz(t) - kz_d(t) \qquad (2.8)$$

Fig. 2.13 Schema of a typical NC-AFM

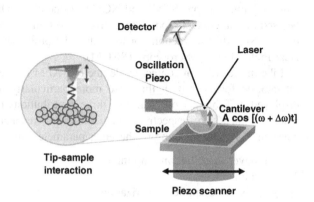

with $z_{d(t)} = A_d \cos(\omega t)$ where A_d is the oscillation amplitude at frequency ω, and m^* the effective mass of the oscillator. The above equation can be also written in the following form:

$$\frac{d^2 z(t)}{dt^2} + \frac{\omega_0}{Q}\frac{dz(t)}{dt} + \omega_0^2 z(t) = A_d \omega_0^2 \cos(\omega t) \tag{2.9}$$

where $\omega_0^2 = K/m^*$ is the resonance frequency of the free oscillator (i.e., $\gamma = 0$), and $Q = m^*\omega_0/\gamma$, the quality factor. It represents the number of oscillation cycles after which the damped oscillation amplitude decays to $1/e$ of the initial amplitude without external oscillation ($A_d = 0$). The steady solution of the equation is reached after $2Q$ cycles of oscillation and displays a phase shift φ:

$$z_1(t) = A_0 \cos(\omega t + \varphi) \tag{2.10}$$

with

$$A_0 = \frac{A_d Q\, \omega_0^2}{\sqrt{\omega^2 \omega_0^2 + Q^2 \left(\omega_0^2 - \omega^2\right)^2}} \tag{2.11}$$

$$\varphi = arctg\left[\frac{\omega\omega_0}{Q\left(\omega_0^2 - \omega^2\right)}\right] \tag{2.12}$$

The amplitude and phase diagrams are displayed in Fig. 2.14 according with these expressions.

The damping term in the oscillator equation of motion causes theoretically a shift in the resonance frequency from ω_0 to ω_0^* but such shift is negligible for $Q > 100$, which is usually obeyed for most of applications in air or vacuum.

Silicon microcantilever tips can be coated for specific purposes, such as a ferromagnetic coating for use as a magnetic force microscope (MFM), or by doping the silicon, the sensor can be made conductive to allow simultaneous scanning tunneling microscopy (STM) and NC-AFM operation. Other oscillator probes can be used in non-contact mode like quartz micro-tuning forks, which are particularly used in NC-AFM in vacuum, or to control the probe-sample distance in Scanning Near-Field Optical Microscopy (SNOM).

Like in the AFM in contact mode, the NC-AFM is sensitive to a combination of interatomic forces, but in this case more particularly to long range forces. As displayed in Table 2.1 one can note three contributions for the frequency shift Δf, where f_0 and K, are respectively the cantilever's resonance frequency and spring constant [68]. The meaning of the other parameters involved are the following:

A—cantilever's oscillation amplitude
R—tip radius
V_{bias}—tip-sample applied polarization

Fig. 2.14 Phase and amplitude versus cantilever excitation frequency for $Q = 4$ in the damped oscillator model [61]

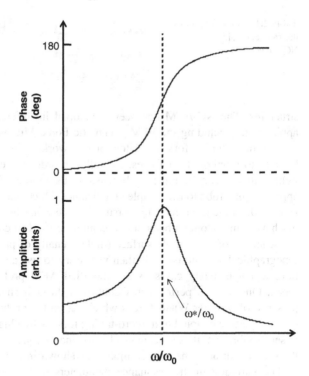

V_{CPD}—contact potential difference
H—Hamaker constant of the tip-sample system
s—tip-sample distance
U_0—Morse potential well depth for the tip-sample system
s_0—Morse potential equilibrium position for the tip-sample system
λ—characteristic length of the interaction

The contact potential difference V_{CPD} is defined by [69]:

$$V_{CPD} = \frac{1}{e}\left(\phi_{tip} - \phi_{sample}\right) \tag{2.13}$$

where ϕ_{tip}, ϕ_{sample} are the work functions of the tip and sample, respectively.

From the three above mentioned contributions, the one that presents the larger sensitivity with the distance s, is the covalent (considering that it is represented by a Morse attractive potential, i.e. the tip is not in the repulsive region of the interatomic potential); this fact favors the recording of high resolution imaging, since the tip approach is kept in the regime where this contribution prevails.

Applying a suitable polarization V_{bias}, it becomes possible to minimize the electrostatic contribution to the frequency shift. This is of a great advantage because the long range electrostatic forces can make difficult the resolution of small

Table 2.1 Contributions for
the frequency shift in
NC-AFM

$\frac{\Delta f_{elec}}{f_0} kA = -\frac{\pi \varepsilon_0 R (V_{bias} - V_{CPD})^2}{\sqrt{2 \bar{s} A}}$	Electrostatic contribution
$\frac{\Delta f_{vdW}}{f_0} kA = -\frac{HR}{12 \bar{s} \sqrt{2 \bar{s} A}}$	van der Waals contribution
$\frac{\Delta f_{CHem}}{f_0} kA = -\frac{U_0}{\sqrt{\pi A}} \sqrt{2} e^{\frac{-s - s_0}{\lambda}}$	Covalent contribution

structures. The NC-AFM has been developed in different varieties for dedicated
applications, including the EFM, Magnetic Force Microscopy (MFM) and KPM.

The micro-tuning fork oscillators probes working in non-contact mode can also
be used to monitor shear forces. Actually, in order to be able to obtain images of
light sources with nanometer dimensions using the SNOM, it is necessary to
approach the probe to the sample at a distance which allows detection of the optical
near field; as the light intensity versus distance from the sample in that region does
not have a monotonous behavior, it cannot be used to control this distance during
the scanning of the sample surface (under penalty of probe destruction due to the
topographical variations). It is then necessary to use an indirect method of detection, which in many cases uses a modified AFM probe running in non-contact
mode. One way to perform this control is through shear force detection on the
presence of Van der Waals forces which act on the probe tip. An example uses an
optical aperture coupled to a micro-tuning fork (with a high quality factor Q) which
is set in vibration at the mechanical resonance frequency in a parallel direction to
the sample surface (shear force mode), as shown in Fig. 2.15.

The variation in the resonance parameters of the oscillations (in general the
phase shift between excitation and oscillation) is used for distance feedback control.

As far as the probe approaches perpendicular to the surface and at a distance of
approximately 20 nm from it, its oscillation amplitude (typically less than 10 nm)
decreases due to a shear force of about 0.1 nN that is present at the probe
tip. A compact way to carry out the monitoring of the distance between the probe
and the sample, is through piezoelectric detection of the interaction probe/sample.
The resonant frequency is measured accurately using a phase sensitive amplifier
(*lock-in*). Once the tuning fork is set to vibrate at its resonant frequency, this signal

Fig. 2.15 Micro-tuning fork
oscillation in shear force
mode

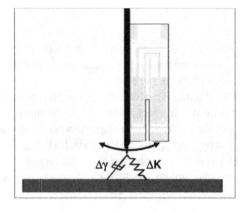

Fig. 2.16 Typical amplitude
and phase response curves of
a micro-tuning fork

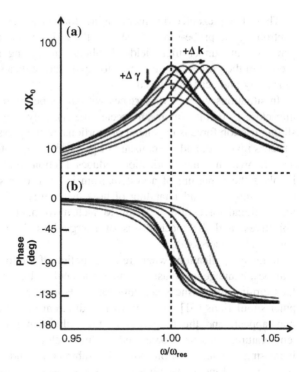

is used to positioning the tip of the optical fiber between 0 and 25 nm above the
sample, using an electronic feedback loop.

Figure 2.16 shows typical response curves of amplitude and phase of a
micro-tuning fork, thus illustrating the atomic interaction force that the optical fiber
tip senses when it oscillates close to the sample surface (in shear force mode). This
force causes the probe to be sensed creating "inertia" and causing a variation in the
amplitude and in the phase of the micro-tuning fork resonance frequency.

The experimental data obtained for the vibration amplitude $z(t)$ of the probe as a
function of the frequency ω, fit very well to the dependence predicted by the forced
harmonic oscillator with damping model, because when the latter is low ($\gamma \ll \omega$),
the differential equation of motion leads to a value of $z(t)$ equal to:

$$\frac{F/m^*}{(\omega_0^2 - \omega^2 + i\gamma\omega)\exp(i\omega t)} \tag{2.14}$$

which translates into a Lorentzian, where $F = A_d \exp(i\omega t)$.

The shear-force for small damping is given by $(im^*\gamma\omega z)$, and in the resonance
(where the amplitude is $z_0 = 3^{1/2}\ FQ/iK$), it depends only on the constant K of the
oscillator, the quality factor Q and z_0. In order to obtain its modulus it is therefore
necessary to determine experimentally the value of these three quantities.

Thus, the piezoelectric micro-tuning fork is an excellent alternative tool for controlling the probe-sample distance, the more it can be carried out at low temperatures and under high fields. It allows measuring the shear-force amplitude present in the probe-sample interaction and represents a significant improvement in sensitivity and scanning speed.

In situations where two surfaces with areas below the micrometer range come into close proximity (as in an accelerometer), they may adhere together. The dry sliding friction force is in general proportional to the pressure force between the two surfaces involved, and the constant of proportionality is the friction coefficient. It is well known that the lubrication reduces friction, and the motion with friction involves the departure of a system from its state of rest to another one far from equilibrium. The understanding of how the energy is irreversibly redistributed as the frictional forces are opposing to such movement is one of the nanotribology objectives, and the mechanisms of energy loss during the sliding are not yet completely understood.

In many applications where relative motion of micro-components takes place, it may be advantageous to use organized molecular films instead of a lubricating fluid, thus turning the surfaces self-lubricating. However, recent experiments and computer simulations [61] have shown that different ordered structures during friction may appear, and the properties of these different states depend mainly on the temperature, pressure force and sliding velocity. With the present trend of increasing miniaturization towards submicron and nanometer scales, these self-organized molecular films can in some cases contribute to increase the friction instead of reducing it.

In fact, the force between two solid surfaces in contact with a liquid or with a self-lubricating molecular film can be attractive, repulsive, oscillatory or even a more complex function of separation; furthermore, when this separation becomes of the order of a few molecular layers, the interface may even solidify [61]. The deep understanding gained in single-molecule friction measurements (including a previously unknown type of friction [70]) opens up new ways to understand nano-friction at the nanometer range, by targeted preparation of new macromolecular ultra-thin films that can be developed specifically for the nanometer range studies.

Using the FFM, some researchers have shown that Amonton's Law (according to which the coefficient of friction is independent of the apparent contact area and normal pressure) does not apply to the sub-microscopic scale. This suggests that the nano-components in relative motion under low contact pressures should experience a small friction, and therefore a negligible wear. In fact, at sub-microscopic and atomic scales, when two surfaces with uniform roughness slide by one another, the phenomenon called "stick-slip" occurs, which is caused by switching between microscopic points of contact and non-contact, with the corresponding variation in the friction force. Indeed, the friction force and normal forces are independent from one another, and this can be understood using the repulsive magnetic analogy with two moving surfaces and under pressure, between which is a confined lubricant film, where frictional forces dependent on the fluid viscosity and velocity slip, even if the forces that push the magnets have nothing to do with these quantities.

Recently, surface-modified carbon nanotubes and other nanocarbons, such as fluorinated nano-onions, polyfullerenes and fluoronanodiamond, both in neat form and as additives to liquid lubricants have been progressively developed [71]. Tribological studies of a series of single-walled carbon nanotubes (SWNT) revealed that the type of chemical treatment of nanotube surface has a significant effect on their lubrication properties, friction coefficients and wear life. The friction coefficients for fluoronanotubes, as well as pristine and chemically cut nanotubes, were found to reach values as low as 0.002–0.07, thus showing a promise for application of surface modified nanocarbons as solid lubricants [71]. Based on combined friction and wear life data, pristine SWNTs and $C_{20}F$ samples show the best lubrication performance in air among all nanotube samples investigated thus far [71]. The trend in SWNT materials as lubricants is partly related to the degree to which they can fill in pits and scratches, and between surface asperities, between sliding surfaces. However, if this were the only mechanism, then all tubes would behave similarly; thus, it must exist some additional effect related with the chemical nature of the surface modifications. A small amount of fluorination can serve to break up SWNT bundles, without significant tube wall degradation, thereby creating a better lubricant, while cut SWNTs and F-cut-SWNTs having similarly tube segments commensurate with surface roughness values, their production alters the graphitic quality of the SWNTs (on the tube ends), leading to a poorer lubricating ability [71]. The magnification of this trend is achieved by the higher fluorination stoichiometry of the C_nF (n = 2, 5, 20) materials, producing a better lubricant by reference to the bulk material, their synthesis leads to substantial SWNT degradation (by introduction of sp^3 hybridization, etc.). Consequently, these highly fluorinated SWNTs can be less resistant towards deformation and more prone to rupture and thus producing dangling bonds and resulting stiction (static friction threshold) far (Fig. 2.17).

In future molecular machines that use mobile elements manufactured with high symmetry diamondoid materials (i.e., pure solids that form covalent tetrahedral bonds, like C, Si, Ge, diamond like …), the viscous and friction forces are likely to be considerably reduced (although not possible to eliminate van der Waals forces). In the Fig. 2.18 two partially nested carbon nanotubes are represented. After relative translation, the inner nanotube undergoes an attractive force that pulls it back to its nested initial position [72]. The concentric geometry configuration of the two nanotubes is maintained by van der Waals forces between the inner and outer tube carbon atoms, and presumably there are also electronic repulsive forces in such a way that a minimum of energy occurs (stable configuration) when the tubes stay in coaxial position. Due to the finite size of the carbon atoms, the interaction potential energy must provide a periodicity, so that the translational and rotational barriers should be small compared to the thermal energy (k_BT).

However, such diamondoid structures do not should be able to assemble atom-by-atom using SPM probes, because the cage of those macromolecules is too small to allow accessibility to objects of the size of SPM probe tips. Moreover, this task would require more than one probe, which further complicates such possibility. Furthermore, to corroborate this impossibility, an atom manipulated by SPM

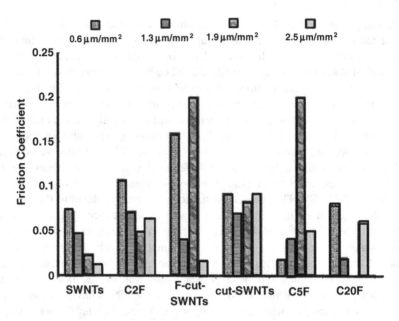

Fig. 2.17 Friction coefficients of SWNT samples determined at different relative coating thickness in contact with sapphire in air [71]

Fig. 2.18 TEM image of partially nested nanotubes after relative translation. Attractive forces pull back the nanotube to its nested initial position. (reprinted with permission [72])

"choose" its preferred connection configuration, instead of the one required by the minimum energy of the overall structure. Since the diamondoid structures have multiple bondings under tension due to its curvature, each atom placed one-by-one in sequence, should choose a bonding without tension. Indeed, the molecular assembler is a subject that has long been debated, but from the point of view of present experimental Nanophysics knowledge, such a device is not yet feasible. On the one hand, a tip is too large to be able to access locations in a complex atomic structure, and on the other hand, a small tip could not orient an atom and adjust its ability to grasp and release the atom. In addition, such device must operate at very low speeds to be able to be used in the production atom-atom appreciable amounts.

In Nature there is nothing that can resembles it, because very specific enzymes catalyze the formation of specific molecules or cut them in certain regions. It is a typical "key-lock" action; instead of acting on atoms, the natural assemblers operate through a large inventory of molecules which are formed or self-organize in accordance with the rules of chemistry. Thus, natural processes are very specific and do not follow a general procedure as the commonly believed molecular assembler. Therefore, if possible the synthetic nano-carriers may pose an alternative to creating new structures that biological methods cannot do. This would be the triumph of physics based nano-engineering as opposed to the spontaneous chemical based biology.

In this line of action, a single molecule designated by "nanocar" was synthesized at Rice University [73], which contains a H-shaped chassis and 4 axis connected to 4 wheels of C_{60} that rotate independently. The fullerene wheels are able to rotate because they bind to the axis of the alkyne via a carbon-carbon single bond. The hydrogen in the neighbor carbon is not an impediment to this rotation. The nanocar moves on a surface by SPM manipulation, in order to judge whether the fullerene wheels rolled or slide. In fact, it has been found that it moves in the direction perpendicular to their axes on a metal surface, rolling and not sliding the four wheels, which is a proof of directional control of its motion [73]. Tests have shown that it is easier to push the nanocar in the direction of rotation of the wheels than in other directions. When the "nanocars" are dispersed over a gold surface, the molecules adsorb onto the gold through the fullerenes, and subsequently observed by STM. Their orientation can be deduced, since the chassis is somewhat smaller than its own width. At room temperature, the adsorption fullerene-gold and the carbon-carbon bond of the fullerene-alkyne do not allow the motion of the nanocar, but by heating at 200 °C, the nanocars start to move. STM images must be obtained at regular time intervals (1 min), and so it is possible to observe the nanocar motion. Once more it was confirmed (now without the aid of STM manipulation) that motion took place in the direction perpendicular to their axis.

Let us now turn slightly on the van der Waals interactions (which in general include three components: permanent dipole-permanent dipole, permanent dipole-induced dipole and induced dipole-induced dipole). At any given instant, any atom or molecule has an asymmetric distribution of charge that corresponds to an instantaneous electrical dipole moment p. The energy of the attractive interaction between a dipole and a non-polar molecule is proportional to r^{-6}. It happens that the so-called van der Waals effect (London dispersion forces) is a dynamic interaction of completely quantum origin, resulting from the fact that the electron in the atom does not have a regular orbit, giving rise to a floating electric dipole. All electrons in an atom or molecule take part in this phenomenon, creating an overall effect equivalent to that of a floating dipole electric field in the surrounding region. The van der Waals effect stems from the fact that this field induce dipoles in polarizable systems such as nearby atoms or molecules. Therefore, atom fluctuations are correlated with the fluctuations in neighboring atoms, resulting in an attractive interaction proportional to r^{-6}.

The calculation of this interaction energy can be obtained by using the quantum mechanical perturbation theory, considering the interaction energy between the atoms with an additional correction that arises from the finite transit time for the propagation of light between the two interacting systems. When the distance between atoms is greater than the distance that light can travel during the characteristic lifetime of the fluctuations, London dispersion forces are weakened. This is the so-called retardation effect, which following Casimir and Polder can be neglected, if such distance is less than 5 nm. If, otherwise, the decay of the attractive force ceases to do with r^{-7}, following then a behavior with r^{-8} (Casimir-Polder equation) [74].

For two many-electron atoms it can be shown that the energy of the van der Waals interaction is [75]:

$$U_{vdW} = -\frac{3}{8\pi\varepsilon_0} \frac{\alpha_A \alpha_B}{r^6} \frac{I_A I_B}{I_A + I_B} \qquad (2.15)$$

where I_A and I_B are the first ionization energies of atoms A and B, and a_A and a_B their respective polarizabilities. Since these polarizabilities are proportional to the number of electrons on each atom, the energy can be hundreds of times greater than the van der Waals interaction between the two hydrogen atoms (which is of the order of 4×10^{-6} eV at a distance of 1 nm, compared with 5 eV for the covalent molecule H_2).

The quantum theoretical calculations have shown that the helium dimer (He_2) should represent the longest and also weaker molecular bond (on the order of 10^{-7} eV). However, the traditional techniques of spectroscopy, as well as diffraction, or electronic scattering have not proved able to detect it, given its extreme fragility. Toennies et al., at MPI Gottingen [76], succeeded in obtaining such experimental evidence, using the molecular beam technique in which cold helium atoms (4.5 K) are isentropically expanded through a nanometric diffraction grid. The peak intensities obtained revealed a bonding length 5.2 nm for He_2 and a bond energy of 9.5×10^{-8} eV, in clear agreement with theoretical expectations.

The van der Waals interaction is attractive and varies with r^{-6}, while the electrostatic interaction varies with r^{-1}. Nevertheless, if an atom or a molecule are interacting with a larger body such as a plane, the dependence of the interaction with the distance to the plane, can be obtained through integration of the interaction energy to all atoms in the plane, assuming additivity of potential pairs of the type V (r) = $-C/r^6$ (which is justified at distances where the three-body interaction can be neglected, i.e. greater than the equilibrium distances). The total interaction energy, U_{vdW}, can thus be estimated by:

$$U_{vdW} = \int_{r_{eq}}^{r} V(r)\rho(r)4\pi r^2 dr \qquad (2.16)$$

where $\rho(r)$ is the number density of the interacting bodies. The output result is proportional to $1/r^3$. Several geometries have been analyzed based on r^{-6} interaction, and some representative results are displayed in Table 2.2 [77].

In these expressions n_v represents the number of atoms per unit of volume and H is the Hamaker constant, that depends on the material properties (it can be positive or negative in sign depending on the intervening medium) being approximately given by:

Table 2.2 Selected van der Waals interaction geometries [86]

$U_{vdW} = -\frac{\pi C n_v}{6r^3}$	Between an atom and a plane separated by a distance r
$U_{vdW} = -\frac{HR}{6r}$	Between a sphere of radius R and a plane at a distance $r \ll R$
$U_{vdW} = -\frac{A}{6}\left(\frac{2R_1 R_2}{z^2-(R_1+R_2)^2} + \frac{2R_1 R_2}{z^2-(R_1-R_2)^2} + \left[\frac{z^2-(R_1+R_2)^2}{z^2-(R_1-R_2)^2}\right]\right)$	Between two spheres of radius R_1, R_2 where $Z = R_1 + R_2 + r$
$U_{vdW} = -\frac{H}{6r}\frac{R_1 R_2}{R_1+R_2}$	Between two spheres of radius R_1, R_2, at a distance $r \ll R_1, R_2$ *note* this equation is obtained from the above one, because in the limit of close-approach the spheres are sufficiently large compared to the distance between them
$U_{vdW} = -\frac{HS}{16\pi r^2}$	Between two parallel planes of area S, separated by a distance r

$$A = \pi^2 C \rho^2 \approx 10^{-19} - 10^{-20} J \qquad (2.17)$$

H is not completely invariable (with a variation between 0.2 and 2.0 eV), due to the fact that constant C is proportional to the square of atomic (or molecular) polarizability, and thus to the square of the volume.

Now one intends to address the Casimir force [78], which operates between two metal surfaces, and becomes significant at very short distance approach. It is an electromagnetic effect originating from the oscillation modes of the electromagnetic field in a cavity (harmonic oscillator model). Two parallel mirrors spaced by a distance L allows the creation of stationary electromagnetic waves propagating in the z direction perpendicular to them, when $L = n\lambda/2$. The state of lowest energy of electromagnetic modes (which behave as nano-oscillators) corresponds to hv/2, as in any harmonic oscillator (zero point energy), which is consistent with the Heisenberg uncertainty principle. Thus, even an empty cavity between mirrors have an energy equal to $Nh\nu/2$, where N is the number of allowed modes in the cavity of width L. The allowed frequencies are given by $nc/(2L)$ where $n = 1, 2, 3, \ldots$, and so frequencies lower than $nc/2L$ are not allowed, i.e. wavelengths longer than $2L$. Therefore, as L decreases, more wavelengths (and more zero point energies) are excluded. This gives rise to an attractive force equal to $-dU/dz$. A recent calculation of the Casimir force, F_c, between parallel mirrors gives the following result [79]:

$$F_c = -\frac{\pi h c}{480}\frac{1}{Z^4} \qquad (2.18)$$

For example, between two 10 nm spaced parallel metal surfaces, it is present an attractive Casimir pressure of about 1 atm. Thus, this force can prove to be very important in future nanomachines with no lubricant. Another simple geometry is a

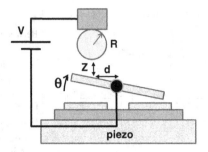

Fig. 2.19 Device (not to scale) for measuring the Casimir force between a sphere and a planar surface measured with V = 0

sphere of radius R at a distance d from a flat surface. In this case, the Casimir force exerted on the ball is [79]:

$$F_{cs} = -\frac{\pi^2 hc}{720} R \frac{1}{d^3} \tag{2.19}$$

This result was fully confirmed using an experimental microdevice [79] illustrated in Fig. 2.19. A plate on which two polycrystalline silicon capacitors are suspended (so that they can be rotated by an angle of torsion θ, through two silicon fibers) approaches a sphere of radius 100 µm (via a piezoelectric stage), at a distance of 75 nm. Both the sphere and the surface of the plates are coated with gold. The 2 µm spacing of the capacitors is obtained by erosion of a SiO_2 layer. The set of suspended capacitors is grown on a single silicon crystal. The force between the ball and the right portion of the blade is determined by the torsion angle θ, which in turn is measured by differential capacitance between the plates. The torsion constant of the fiber is calibrated through the torsion angle observed by using a bias applied to the ball, at a distance such that the Casimir force is negligible. The Coulomb force between the sphere of radius R and the plane at a distance z is a known function of R, V, z, according to which the measures were carefully adjusted, and so calibrating this way the force versus torsion angle θ. The forces measured versus z spacing are shown in Fig. 2.20.

One can explain the spacing between the theoretical and experimental curves based on two factors. On the one hand, the films of metallic gold deposited on surfaces are not perfect mirrors, thereby allowing high frequency electromagnetic waves passing through them. This effect can be corrected using tabulated optical constants for gold. The second effect is that gold surfaces are not perfectly flat, having a surface roughness amplitude of 30 nm, which is measured by SPM.

Placing two perfect mirrors facing each other at a very short distance in empty space would disturb the zero-point electromagnetic field, because in such a region, only the wavelengths that fit exactly into the space between the mirrors are allowed. If only certain wavelengths of real photons are allowed, then something must have happened to the zero-point field that forbids it to generate other wavelengths of

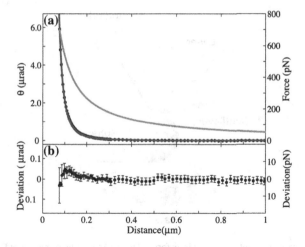

Fig. 2.20 Measured Casimir force between a sphere and a plane versus spacing z. **a** Lower trace: data points are fitted to theoretical curve; upper trace: force versus z for V = 136 mV where Coulomb force matches Casimir force at 76 nm (closest approach). **b** Expanded scale for the deviation between data points and fitted theoretical Casimir curve. (reprinted with permission [79])

photon. The zero-point energy density must therefore be lower in the region between the mirrors than outside, and this difference in energy will depend on the distance between the mirrors. An energy that depends on distance equals a force, and so Casimir's simple prediction was that two mirrors in empty space would attract each other [80]. The force can be quite large at very small distances: for a 10 nm separation of the mirrors, the pressure on them is the same size as atmospheric pressure (10^5 N/m^2). However, it drops very rapidly (as the inverse 4th power of separation for perfect mirrors), and so it has been very difficult to accurately measure such force between two sufficiently flat and sufficiently smooth parallel surfaces approaching from micron to submicron distances.

Casimir's theory made the assumption that the mirrors were perfect (i.e., totally reflecting at all wavelengths) and it was extended by Lifshitz [81] to include real mirrors made out of real materials. In 1997 Lamoreaux measured the Casimir force between a metal sphere and a metal plate using a very sensitive torsion balance [82]. The sphere-plate geometry can still be rigorously calculated and produces a smaller Casimir force at a given separation, with the advantage of getting rid of the huge experimental difficulty that is keeping perfect parallelism down to tiny separations. For perfect parallel reflectors the Casimir force depends only on the area of the plates, A, the distance between them, d, and the universal constants, c (the speed of light) and h (Planck's constant). In the sphere-plate geometry the Casimir force depends only on the radius R of the sphere, the gap between the sphere and the plate, d, and the same universal constants. The magnitude of the force, for a given value of d is less in the sphere-plate configuration, but the experimental problem of maintaining perfect parallelism at sub-micron separations is largely removed.

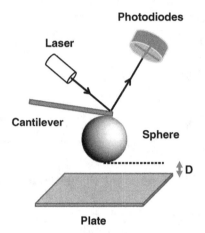

Fig. 2.21 AFM measurement of the Casimir force. The AFM cantilever has a gold-coated sphere glued onto the end instead of having a sharp tip as normally used to scan a surface topography. The deflection of the cantilever is determined in the usual way

The utilization of a modified AFM in Casimir force measurements was pioneered by Mohideen et al. [83], using a 200 μm diameter gold-coated polystyrene sphere, and has then become of great practical significance in micro- and nanoscale mechanical devices (Fig. 2.21). As the size of these devices has decreased, they have become full of boundaries with submicron gaps where the Casimir force becomes dominant. In fact, it is a significant problem because while one can take measures to prevent things like capillary and electrostatic forces, there is nothing that can be done to prevent the Casimir force as it arises from the fundamental properties of the vacuum. It thus generates a fundamental and ever-present stickiness of components in micromachines and nanomachines.

Casimir force has also the potential to transmit force at short distances without physical contact, and Capasso et al., have been successful in modifying the motion in a micromechanical system [84]. They built a standard micromechanical device consisting of a flat silicon plate with dimensions of a few hundred microns, which is suspended by a torsion wire above a surface (Fig. 2.22). By applying an AC voltage to the pads underneath the plate, it can be made to oscillate in seesaw fashion at a frequency of a few kHz. Then, using an AFM-type manipulator, they lowered a 100 μm gold-coated sphere toward one side of the oscillating plate approaching it to within a few hundred nanometers. The Casimir force between the sphere and the plate causes a shift in the frequency of the oscillator. They found that the amplitude and frequency shift of the oscillator were measurable with only a few nanometers change in the height of the sphere [79].

Fig. 2.22 Utilization of the Casimir force to modify the frequency of a micro-mechanical oscillator by using AFM technology to bring a gold-coated sphere to within a few hundred nanometers of one side of a torsion oscillator

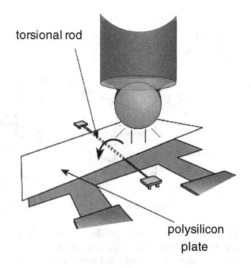

torsional rod

polysilicon plate

More recently, the Mohideen group also demonstrated the existence of a lateral force between a corrugated surface and a similar corrugation imprinted onto a gold-coated sphere [85] (Fig. 2.23). This way, when the sphere is moved parallel to the corrugation, a lateral force that tends to drag the corrugated surface in the same direction is generated through the vacuum.

This is likely to contribute for the stickiness problem in forthcoming nanomachines, and as an illustration one addresses the basic idea for a simple rack and pinion shown in Fig. 2.24. A rack is moved laterally at a distance of a few tens to hundreds of nanometers away from a pinion, and a force is transmitted through the vacuum to rotate the pinion. It must be find the right combination of materials and the right shape of corrugations to make this kind of machine a practical reality.

It is likely that the strength of the force transmission and even the direction of the force depend on the speed at which the rack is moved. Thus if it is moved backwards and forwards at different speeds in each direction, it may be possible to provide a continuous unidirectional torque on the pinion. This is mechanically useful for the transfer of oscillatory motion into a unidirectional linear motion (analogy with a car transmission system from the oscillating pistons to the linear driving motion).

An alternative method of achieving that is to define a system with an asymmetric set of teeth as shown in Fig. 2.25. Then oscillating the two surfaces in the normal direction will provide a unidirectional linear force, whose direction can be varied by altering the lateral displacement of the two sets of teeth. This is the so-called Casimir ratchet effect.

A special emphasis should be addressed to the distinction between the Casimir and van der Waals forces since they both have origin in the same source: the zero-point electromagnetic field. Their theoretical descriptions are different, because in general the van der Waals force is expressed in terms of surface charges on the

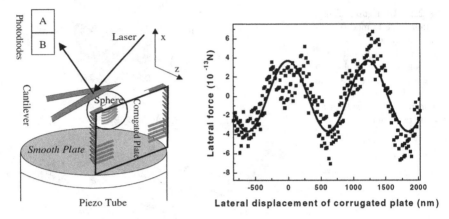

Fig. 2.23 Demonstration of the lateral Casimir force transmitted trough vacuum between a corrugated surface and a sphere with the same corrugation imprinted onto it. Pulling it past the surface generates a lateral force that oscillates as the corrugation move past each other. (reprinted with permission [85])

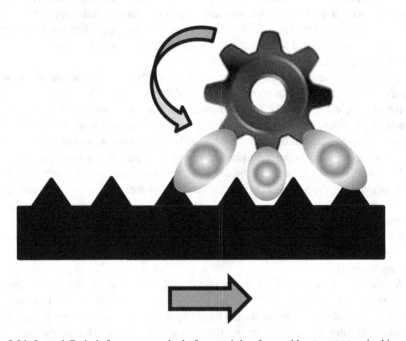

Fig. 2.24 Lateral Casimir force as a method of transmitting force without contact as in this rack and pinion

Fig. 2.25 Casimir ratchet

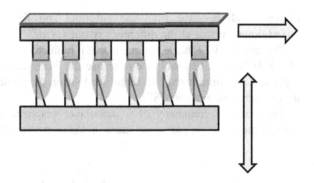

two bodies whereas the Casimir force is described in terms of the zero-point electromagnetic field. The surface charges, however, are a result of fluctuations in the zero-point field; hence fundamentally the two forces arise from the same source. As two surfaces are brought together, the power law describing how the force varies with distance changes, and this is the crossover between the two regimes. This crossover can be thought of as the minimum distance between two components before they stick together. This distance is about 10 nm, and so the rack and pinion machine shown in Fig. 2.25 could be scaled down until the teeth were about the size of large molecules.

Finally, let us draw some more considerations regarding nano-resonators and nanomotors. Actually, most of the current NEMS devices fabricated on atomically thin graphene are nano-resonators. Understanding damping (energy loss) mechanisms on thin graphene structures is a first step for applications in radiofrequency, ultrasensitive mass sensing and quantum mechanical information processing. Such nano-resonators are suspended graphene sheets made by etching of the silicon oxide sacrificial layer or by the transfer of graphene to a pre-patterned substrate [87]. These suspended beams or sheets are actuated either electrostatically or optically [88] and their resonant frequencies and quality factors are measured.

In general, graphene suspended nanostructures are fabricated by etching the underlying oxide using buffered hydrofluoric acid or the transfer of graphene sheets to a substrate where trenches are pre-made [88]. In the first method, graphene layers are prepared by exfoliation of graphite and transferred to a substrate coated with a nano-film of silicon oxide. In the second method, trenches are pre-patterned onto the underlying oxide and contact electrodes are deposited and then exfoliated or CVD-grown graphene is transferred and suspended over the trenches.

To measure the resonant frequency and the quality factor of graphene resonators, optical and electrical methods can be used. The high-frequency mixing approach is an electrical measurement technique developed to detect the motion of nanotube resonators [89]. In this method a DC voltage gate V_g is applied to the device in combination with a radiofrequency gate voltage of frequency f that drives the motion [$V = V_g + a \cos(2\pi ft)$]. A second rf voltage, at a frequency $f + \Delta f$, is applied to the source ($\Delta f / f \ll 1$). During the motion of the resonator, its conductance changes with distance from the substrate and the motion is detected as a mixed-down current at the difference frequency Δf.

The drawback of this technique is that it limits the measurements to a bandwidth of 1 kHz [87]. A better choice is to use the Xu et al., technique where the stray capacitance is minimized and direct readout on graphene resonators is performed using a network analyzer [87].

The resonant frequency of a monolayer graphene depends on the physical state of the beam, in particular strain and absorbed mass, and can be modeled by [88]:

$$f = \frac{k}{2}\sqrt{\frac{Y}{\rho_0 L^2}}\sqrt{\frac{s}{\alpha}} \tag{2.20}$$

where L is the graphene beam length, Y and ρ_0 are the Young modulus and the mass density of monolayer graphene, respectively, $k = 1, 2, ..., s$ is the in-plane strain and α is the absorbed mass factor, i.e. $\alpha = \rho/\rho_0$ where ρ represents the density including absorbed mass.

This 1/L dependence on the natural frequency of graphene beam monolayer has been confirmed experimentally [88]. The resonant frequency of monolayer graphene resonator also depends on the gate voltage, which on its turn varies with temperature and the quality factor around 5 K is 14,000 [88].

When a small amount of mass δm is added to a resonating nanobeam, a shift in its resonance frequency is observed. To obtain this frequency shift, one makes use of the Rayleigh approximation, where the beam is treated as a harmonic oscillator with a mass m and a stiffness k. The equivalent mass and the stiffness are given by the following relations [90]:

$$m = \int_0^L \rho A \emptyset(x)^2 dx \tag{2.21}$$

$$k = \int_0^L YI[\emptyset(x)^n]^2 dx \tag{2.22}$$

where $\phi(x)$ is a shape function satisfying the beam differential equation and the appropriate boundary condition; ρ, Y, I and A are respectively, the mass density of the beam, the Young modulus, the area moment of inertia and the beam cross-section. If one neglects the electrostatic spring softening (which appears during electrostatic actuation) and assume that the added mass does not change the spring constant of the beam, then the approximate resonant frequency is given by:

$$f_0 = \frac{\sqrt{k}}{2\pi}\frac{1}{\sqrt{m}} \tag{2.23}$$

Therefore, it is possible to calculate the resonance frequency shift δf for a small variation of mass δm:

$$\delta f = f_0 - f' = f_0 \left(1 - \frac{1}{\sqrt{1 + \frac{\delta m}{m}}} \right) \approx f_0 \frac{\delta m}{2m} = R \delta m \, (\delta m \ll m) \qquad (2.24)$$

where the responsivity R is defined as $f_0/2m$. Consequently, in order to increase the beam responsivity, a small mass and a high resonant frequency are desired. Such requirement can be achieved with nanotubes as their mass is extremely low ($\sim 10^{-21}$ kg) and their mechanical properties allow for very high resonant frequencies. For instance, carbon nanotube mass sensors achieving atomic resolution have been fabricated (i.e. a responsivity of 0.104 MHz zg^{-1}, which corresponds to a few tens of gold atoms) [91]. These devices displayed a sensitivity of 0.4 gold atoms/Hz$^{1/2}$ which is extremely low considering that the measurements are performed at room temperature. Beams clamped at one end have a larger travel range when bent than the travel range of clamped-clamped beams, allowing increased dynamic range. Another advantage of using cantilevers is that the loss owing to clamping is less than with doubly clamped ones.

The performance and reliability of MEMS/NEMS depend critically on the ability of the contacting surfaces to remain as conductive as possible. Surface contamination is a known cause of MEMS/NEMS switches failure due to retarding of current flow [92]. Critical point CO_2 drying can be apply to MEMS/NEMS and combined TEM and Raman investigations on individual SWNTs (in particular a SWNT clamped between two nanowires). At the critical temperature the densities of the liquid and gas become equal (the boundary disappears), and so it will fill the container like a gas, but may be much denser than a typical gas (it is called a supercritical fluid). Carbon dioxide liquefies under pressure at room temperature and above 31 °C no amount of pressure will liquefy CO_2 (this is the critical temperature, T_c).

A key application of NEMS is atomic force microscope tips. The increased sensitivity achieved by NEMS leads to smaller and more efficient sensors to detect stresses, vibrations, forces at the atomic level, and chemical signals.

Thanks to their excellent mechanical properties as well as interesting electrical characteristics, carbon nanotubes are among the most widely used materials for the study of electromechanical properties. Many of the commonly used materials for NEMS technology have been carbon based, specifically diamond, carbon nanotubes and graphene. The large Young's modulus of carbon is fundamental to the stability of NEMS while the metallic and semiconductor conductivities of carbon based materials allow them to function as transistors.

Both graphene and diamond exhibit high Young's modulus, low density, low friction, excessively low mechanical dissipation, and large surface area. The low friction of CNTs, allow practically frictionless bearings and has thus been a huge motivation towards practical applications of CNTs as constitutive elements in NEMS, such as nanomotors, switches and high-frequency oscillators. Carbon nanotubes and graphene's physical strength allows carbon based materials to meet higher stress demands, and thus further support their use as a major materials in NEMS technology.

Along with the mechanical benefits of carbon based materials, the electric properties of carbon nanotubes and graphene allow them to be used in many electrical components of NEMS. Nanotransistors have been developed for both carbon nanotubes as well as graphene.. Transistors are one of the basic building blocks for all electronic devices, and so by developing usable transistors, CNTs and graphene are both very crucial to NEMS.

Metallic CNTs have also been proposed for nanoelectronic interconnects since they can carry high current densities. This is also a very useful property as wires to transfer current are another basic building block of any electrical system. CNTs have found so much use in NEMS that methods have already been discovered to connect suspended CNTs to other nanostructures. This allows CNTs to be structurally set up to make complicated nanoelectric systems.

The rapid miniaturization of devices and machines boost the evolution of advanced fabrication techniques. However, the complexity and high cost of the high-resolution lithographic systems are motivating unconventional routes for nanoscale patterning. Inspired by natural nanomachines, synthetic nanomotors have recently demonstrated remarkable performance and functionality. Nanomotor lithography, is a new nano-patterning approach, which translates the autonomous motion trajectories of nanomotors into controlled surface features. Metallic nanowire motors as mobile nanomasks and Janus sphere motors as near-field nanolenses to manipulate light beams for generating a myriad of nanoscale features through modular nanomotor design have been used [93]. The complex spatially defined nanofeatures using these dynamic nanoscale optical elements can be achieved through organized assembly and remote guidance of multiple nanomotors. Such ability to transform predetermined paths of moving nanomachines to defined surface patterns provides a unique nanofabrication platform for creating diverse nanodevices.

The development of rotary nanomotors is crucial for advancing nano-electromechanical system technology. Nanomotors can be bottom-up assembled from nanoscale building blocks with nanowires as rotors, patterned nanomagnets as bearings and quadrupole microelectrodes as stators. Arrays of nanomotors rotate with controlled angle, speed (over 18,000 r.p.m.), and chirality by electric fields [94]. Using analytical modelling, it has been revealed that fundamental nanoscale electrical, mechanical and magnetic interactions in the nanomotor system, excellently agrees with experimental results and provides critical understanding for designing metallic nano-electromechanical systems. With all its dimensions under 1 μm in size, it is capable of rotating for 15 continuous hours at a speed of 18,000 rpm, representing an important milestone toward developing miniature machines at the sub-micron level [94]. In the experiments, researchers used the technique to turn the nanomotors on/off and propel the rotation either clockwise or counterclockwise. They also found that they could position the motors in a pattern and move them in a synchronized fashion; this makes nanomotors more powerful and with a enhanced flexibility [94].

References

1. M.S. Whittingham, Materials challenges facing electrical energy storage. MRS Bull. **33**, 411 (2008)
2. J. Gomez, L. Vazquez, A.M. Baro, N. Garcia, C. Perdriel, W.E. Triaca, A.J. Arvia, Surface-topography of (100)-type electro-faceted platinum from scanning tunneling microscopy and electrochemistry. Nature **323**, 612 (1986)
3. L. Vazquez, J. Gomez, A.M. Baro, N. Garcia, M.L. Marcos, J. Gonzalez Velasco, J.M. Vara, A.J. Arvia, J. Presa, Scanning tunneling microscopy of electrochemically activated platinum surfaces—a direct ex-situ determination of the electrode nanotopography. J. Am. Chem. Soc. **109**, 1730 (1987)
4. J. Gomez, L. Vazquez, A.M. Baro, C. Alonso, E. González, J. González-Velasco, A.J. Arvia, Scanning tunneling microscopy and scanning electron-microscopy of electrodispersed gold electrodes. J. Electroanal. Chem. **240**, 77 (1988)
5. H. Siegenthaler, *Scanning Tunneling Microscopy, Springer Series in Surface Science*, vol. 28 (Springer Press, Berlin, 1992)
6. C.A. Melendres, A. Tadjeddine (eds.), *Synchrotron Techniques in Interfacial Electrochemistry*, 1st edn. NATO Science Series C, vol. 432. (Springer, London, 1994)
7. R. Sonnenfeld, P.K. Hansma, Atomic-resolution microscopy in water. Science **232**, 211 (1986)
8. O. Lev, F.R. Fan, A.J. Bard, The Application of scanning tunneling microscopy to in-situ studies of nickel electrodes under potential control. J. Electrochem. Soc. **135**, 783 (1988)
9. K. Itaya, E. Tomita, Scanning tunneling microscope for electrochemistry—a new concept for the in-situ scanning tunneling microscope in electrolyte-solutions. Surf. Sci. **201**, 507 (1988)
10. P. Lustenberger, H. Rohrer, R. Christoph, H. Siegenthaler, Scanning tunneling microscopy at potential controlled electrode surfaces in electrolytic environment. J. Electroanal. Chem. **243**, 225 (1988)
11. A.J. Bard, J. Faulkner, *Electrochemical Methods: Fundamentals and Applications*, 2nd edn. (Wiley, New York, 2001)
12. T.R. Cataldi, I.G. Blackham, G. Andrew, D. Briggs, J.P. Pethica, H. Allen, O. Hill, In-situ scanning tunneling microscopy—new insight for electrochemical electrode/surface investigations. J. Electroanal. Chem. **290**, 1 (1990)
13. H. Siegenthaler, STM in electrochemistry, in *Scanning Tunneling Microscopy II*, ed. by R. Wisendanger, H. J. Guntherodt (Springer, New York, 1992)
14. R. Christoph, H. Siegenthaler, H. Rohrer, H. Wiese, In-situ scanning tunneling microscopy at potential controlled Ag(100) substrates. Electrochimica Acta **34**, 1011 (1989)
15. J. Vetter, P. Novak, M.R. Wagner, C. Veit, K.C. Moller, J.O. Besenhard, M. Winter, M. Wohlfahrt-Mehrens, C. Vogler, A. Hammouche, Ageing mechanisms in lithium-ion batteries. J. Power Sour. **147**, 269 (2005)

© The Author(s) 2015
R.F.M. Lobo, *Nanophysics for Energy Efficiency*,
SpringerBriefs in Energy, DOI 10.1007/978-3-319-17007-7

16. E. Peled, The electrochemical behavior of alkali and alkaline earth metals in nonaqueous battery systems—the solid electrolyte interphase model. J. Electrochem. Soc. **126**, 2047 (1979)

17. S.C. Nagpure, R. Dinwiddie, S.S. Babu, G. Rizzoni, B. Bhushan, T. Frech, Thermal diffusivity study of aged li-ion batteries using flash method. J. Power Sour. **195**, 872 (2010)

18. H.E. Exner, E. Arzt, Sintering processes, Ch. 30, in *Physical Metallurgy*, 3rd edn., ed. by R.W. Cahn, P. Haasen (Elsevier Science Publication, 1983)

19. A. Bhattacharya, V.V. Calmidi, R.L. Mahajan, Thermophysical properties of high porosity metal foams. Int. J. Heat Mass Transfer. **45**, 1017 (2002)

20. B. Bhushan, *Nanotribology and Nanomechanics: An Introduction*, 2nd edn. (Springer, Heidelberg, 2008)

21. S.C. Nagpure, B. Bhushan, S.S. Babu, G. Rizzoni, Scanning Spreading resistance characterization of aged Li-ion batteries using atomic force microscopy. Scripta Mater. **60**, 933936 (2009)

22. S.C. Nagpure, S.S. Babu, B. Bhushan, A. Kumar, R. Mishra, W. Windl, L. Kovarik, M. Mills, Local electronic structure of $LiFePO_4$ nanoparticles in aged Li-ion batteries. Acta Mater. **59**, 6917 (2011)

23. P. Eyben, T. Janssens, W. Vandervorst, Scanning spreading resistance microscopy (SSRM) 2d carrier profiling for ultra-shallow junction characterization in deep-submicron technologies. Mater. Sci. Eng. B **124**, 45 (2005)

24. K. Slater, Scanning spreading resistance microscopy (SSRM), in *Support Note No. 294 Rev. A, Veeco Instruments Inc., Santa Barbara, CA* (2000)

25. Anonymous, *Application Modules Training Supplement* (Veeco Instruments Inc., Santa Barbara, CA, 2004)

26. S. Franger, F. Le Cras, C. Bourbon, H. Rouault, Comparison between different $LiFePO_4$ synthesis routes and their influence on its physico-chemical properties. J. Power Sour. **119**, 252 (2008)

27. J.B. Goodenough, Cathode materials: a personal perspective. J. Power Sour. **174**, 9961000 (2007)

28. S.C. Nagpure, S.S. Babu, B. Bhushan, Surface potential measurement of aged Li-ion batteries using Kelvin probe microscopy. J. Power Sour. **196**, 1508 (2011)

29. A.L. Zharin, D.A. Rigney, Application of the contact potential difference technique for on-line rubbing surface monitoring. Tribol. Lett. **4**, 205 (1998)

30. A. Rice, *Nanoscope Controller Manual* (Rev. B, Veeco Instruments Inc., Santa Barbara, CA, 2002)

31. B. Bhushan, A. Goldade, Measurements and analysis of surface potential change during wear of single-crystal silicon (100) at ultralow loads using Kelvin probe microscopy. Appl. Surf. Sci. **157**, 373 (2000)

32. A. Yamada, S.C. Chung, K. Hinokuma, Optimized $LiFePO_4$ for lithium battery cathodes. J. Electrochem. Soc. **148**, A224 (2001)

33. J.D. Wilcox, M.M. Doeff, M. Marcinek, R. Kostecki, Factors influencing the quality of carbon coatings on $LiFePO_4$. J. Electrochem. Soc. **154**, A389 (2007)

34. C.M. Julien, K. Zaghib, A. Mauger, M. Massot, A. Ait-Salah, M. Selmane, F. Gendron, Characterization of the carbon coating onto $LiFePO_4$ particles used in lithium batteries. J. Appl. Phys. **100**, 63511 (2006)

35. M.M. Doeff, Y. Hu, F. McLarnon, R. Kostecki, Effect of surface carbon structure on the electrochemical performance of $LiFePO_4$. Electrochem. Solid St. **6**, A207 (2003)

36. E. Cubukcu, E. Kort, K. Crozier, F. Capasso, Plasmonic laser antenna. Appl. Phys. Lett. **89**, 093120 (2006)

37. N. Yu, E. Cubukcu, L. Diehl, D. Bour, S. Corzine, J. Zhu, G. Hofler, K. Crozier, F. Capasso, Bowtie plasmonic quantum cascade laser antenna. Opt. Express **15**, 13272 (2007)

38. T. Taminiau, R. Moerland, F. Segerink, L. Kuipers, N. van Hulst, l/4 resonance of an optical monopole antenna probed by single molecule fluorescence. Nano Lett. **7**, 28 (2007)

39. R. Lobo, M. Costa, J. Ribeiro, C. Sequeira, P. Pereira, Fluorescence in nanostructured fulleride films. Appl. Phys. Lett. **89**, 203102 (2006)
40. O.V. Kuznetsov, M.X. Pulikkathara, R. Lobo, V. Khabashesku, Solubilization of carbon nanoparticles, nanotubes, nano-onions, and nanodiamonds through covalent functionalization with sucrose. Russ. Chem. Bull. **59**, 1495 (2010). International Edition (Springer)
41. L. Zhang, V. Kiny, H. Peng, R. Lobo, J. Margrave, V. Khabashesku, Sidewall functionalization of single-walled carbon nanotubes with hydroxyl group-terminated moieties. Chem. Mater. **16**, 2055 (2004)
42. R. Lobo, F. Berardo, J. Ribeiro, In-situ monitoring of lowermost amounts of hydrogen desorbed from materials. Int. J. Hydrog. Energy **35**, 11405 (2010)
43. R. Lobo, F. Berardo, J. Ribeiro, Molecular beam-thermal hydrogen desorption from palladium. Rev. Sci. Instrum. **81**, 43103 (2010)
44. N. Yao, Z. Wang, *Handbook of Microscopy for Nanotechnology* (Kluwer Acad. Publ, London, 2005)
45. U. Weierstall, J. Spence, Atomic species identification by STM using an imaging atom-probe technique. Surf. Sci. **398**, 267 (1998)
46. T. Shimizu, J. Kim, H. Tokumoto, Tungsten silicide formation on an STM tip during atom manipulation. Appl. Phys A **66**, S771 (1998)
47. A. Kumar, F. Ciucci, A.N. Morozovska, S.V. Kalinin, S. Jesse, Measuring oxygen reduction/evolution reactions on the nanoscale. Nat. Chem. **3**, 707 (2011)
48. M.K. Song, S. Park, F. Alamgir, J. Cho, M. Liu, Nanostructured electrodes for lithium-ion and lithium-air batteries. Mater. Sci. Eng. Rep. **72**, 203 (2011)
49. A.T. D'Agostino, W.N. Hansen, Observation of systematic electrochemically induced binding energy shift in the XPS spectra of emersed Cs^+ double layer species. Surf. Sci. **165**, 268 (1986)
50. W.N. Hansen, The emersed double layer. J. Electroanal. Chem. **150**, 133 (1983)
51. D. Kolb, D. Rath, R. Wille, W. Hansen, An ESCA study on the electrochemical double layer of emersed electrodes. Berichte de Bunsengesellschaft/Phys. Chem. Chem. Phys. **87**, 1108 (1983)
52. A. Foelske-Schmitz, D. Weingarth, R. Klotz, Quasi in situ XPS study of electrochemical oxidation and reduction of highly oriented pyrolytic graphite in [1-ethyl-3-methylimidazolium][BF$_4$] electrolytes. Electrochim. Acta **56**, 10321 (2011)
53. R. Kotz, H. Neff, S. Stucki, Anodic iridium oxide films XPS-studies of oxidation state changes and O_2 evolution. J. Electrochem. Soc. **131**, 72 (1984)
54. A. Foelske, H.H. Strehblow, Structure and composition of electrochemically prepared oxide layers on Co in alkaline solutions studied by XPS. Surf. Interf. Anal. **34**, 125 (2002)
55. M. Beerbom, T. Mayer, W. Jaegermann, Synchrotron-Induced Photoemission of Emersed GaAs Electrodes after Electrochemical Etching in Br_2/H_2O Solutions. J. Phys. Chem. B **104**, 8503 (2000)
56. W. Zhou, D.M. Kolb, Influence of an electrostatic potential at the metal/electrolyte interface on the electron binding energy of adsorbates as probed by X-ray photoelectron spectroscopy. Surf. Sci. **573**, 176 (2004)
57. A.W. Taylor, F. Qiu, I.J. Villar-Garcia, P. Licence, Spectroelectrochemistry at ultrahigh vacuum: in-situ monitoring of electrochemically generated species by X-ray photoelectron spectroscopy. Chem. Commun. **39**, 5817 (2009)
58. D. Gentili, M. Cavallini, Wet-lithographic processing of coordination compounds. Coord. Chem. Rev. **257**, 2456 (2013)
59. R.V. Martinez, J. Martinez, M. Chiesa, R. Garcia, E. Coronado, E. Pinilla-Cienfuegos, S. Tatay, Large-scale nanopatterning of single proteins used as carriers of magnetic nanoparticles. Adv. Mater. **22**, 588 (2010)
60. H.P. Lang, M. Hegner, E. Meyer, Ch. Gerber, Nanomechanics from atomic resolution to molecular recognition based on atomic force microscopy technology. Nanotechnology **13**, R29 (2002)
61. B. Bhushan, *Handbook of Nanotechnology*, 1st edn. (Springer, Heidelberg, 2004)

62. J. Ruan, B. Bhushan, Atomic-scale friction measurements using friction force microscopy. J. Tribol. **116**, 378 (1994)
63. B. Bhushan, *Principles and Applications of Tribology*, 2nd edn., 2013. (Wiley, New York, 1999)
64. B. Bhushan, H. Fuchs, S. Hosaka, *Applied scanning probe methods* (Springer, Heidelberg, 2004)
65. P. Reimann, M. Evstigneev, Description of atomic friction as forced Brownian motion. New J. Phys. **7**, 25 (2005)
66. T. Bouhacina, J. Aimé, S. Gauthier, D. Michel, V. Heroguez, Tribological behavior of a polymer grafted on silanized silica probed with a nanotip. Phys. Rev. B **56**, 7694 (1997)
67. E. Gnecco, R. Bennewitz, T. Gyalog, Ch. Loppacher, M. Bammerlin, E. Meyer, H. Güntherodt, Velocity dependence of atomic friction. Phys. Rev. Lett. **84**, 1172 (2000)
68. F.J. Giessibl, Advances in atomic force microscopy. Rev. Modern Phys. **75**, 949 (2003)
69. L. Gross, F. Mohn, N. Moll, P. Lijeroth, G. Meyer, The chemical structure of a molecule resolved by atomic force microscopy. Science **325**, 1110 (2009)
70. B. Balzer, M. Gallei, M. Hauf, M. Stalhofer, L. Wiegleb, A. Holleitner, M. Rehahn, T. Hugel, Nanoscale friction mechanisms at solid-liquid interfaces. Ang. Chem. Int. Ed. **52**, 6541 (2013)
71. R. Wal, K. Miyoshi, A. Tomasek, Y. Liu, J. Margrave, V. Khabashesku, Friction properties of surface-fluorinated carbon nanotubes. Wear **259**, 738 (2005)
72. J. Cummings, A. Zettl, Low-friction nanoscale linear bearing realized from multiwall carbon nanotubes. Science **289**, 602 (2000)
73. Y. Shirai, A.J. Osgood, Y. Zhao, K.F. Kelly, J.M. Tour, Directional control in thermally driven single-molecule nanocars. Nano Lett. **5**, 2330 (2005)
74. R.L. Jaffe, The casimir effect and the quantum vacuum. Phys. Rev. D **72**, 21301 (2005)
75. K.S. Krane, *Modern Physics*, 2nd edn. (Wiley, New York, 1995)
76. R. Grisenti, W. Scholkopf, J. Toennies, G. Kohler, M. Stoll, Determination of the bond length and binding energy of the helium dimer by diffraction from a transmission grating. Phys. Rev. Lett. **85**, 2284 (2000)
77. J.N. Israelachevili, *Intermolecular and surface forces* (Academic Press, San Diego, 1992)
78. H.B.G. Casimir, On the attraction between two perfectly conducting plates. Proc. R. Neth. Acad. Arts Sci. B **51**, 793 (1948)
79. H.B. Chan, V.A. Aksyuk, R.N. Kleiman, D.J. Bishop, F. Capasso, Quantum mechanical actuation of microelectromechanical systems by the casimir force. Science **291**, 1941 (2001)
80. P. Milonni, *The Quantum Vacuum* (Academic Press, San Diego, 1994)
81. E.M. Lifshitz, Theory of molecular attractive forces between solids. Sov. Phys. JETP **2**, 73 (1956)
82. S.K. Lamoreaux, Demonstration of the Casimir force in the 0.6 μm to 6 μm range. Phys. Rev. Lett. **78**, 5 (1997)
83. U. Mohideen, A. Roy, Precision measurement of the Casimir force from 0.1 to 0.9 μm. Phys. Rev. Lett. **21**, 4549 (1998)
84. H.B. Chan, V.A. Aksyuk, R.N. Kleiman, D.J. Bishop, F. Capasso, Nonlinear micromechanical Casimir oscillator. Phys. Rev. Lett. **87**, 211801 (2001)
85. F. Chen, U. Mohideen, Demonstration of the lateral Casimir force. Phys. Rev. Lett. **88**, 101801 (2002)
86. H. Hamaker, The London-van der Waals attraction between spherical particles. Physica **4**, 1058 (1937)
87. Y. Xu, C. Chen, V. Deshpande, F. DiRenno, A. Gondarenko, D. Heinz, S. Liu, P. Kim, J. Hone, Radio frequency electrical transduction of graphene mechanical resonators. Appl. Phys. Lett. **97**, 243111 (2010)
88. A. van der Zande, R. Barton, J. Alden, C. Ruiz-Vargas, W. Whitney, P. Pham, J. Park, J. Parpia, H. Craighead, P. McEuen, Large-scale arrays of single-layer graphene resonators. Nano Lett. **10**, 4869 (2010)

89. B. Witcamp, M. Poot, H. van der Zant, Bending-mode vibration of a suspended nanotube resonator. Nano Lett. **6**, 2904 (2006)
90. W. Stokey, Vibration of systems having distributed mass and elasticity, in *Shock and Vibration Handbook*, 4th edn. ed. by C. Harris (McGraw-Hill, New York, 1988)
91. K. Jensen, K. Kim, A. Zettl, An atomic-resolution nanomechanical mass sensor. Nat. Nanotechnol. **3**, 533 (2008)
92. J. Schimkat, Contact measurements providing basic design data for micro-relay actuators. Sens. Actuators A **73**, 138 (1999)
93. J. Li, W. Gao, R. Dong, A. Pei, S. Sattayasamitsathit, J. Wang, Nanomotor lithography. Nat. Commun. **5**, 5026 (2014)
94. K. Kim, X. Xu, J. Guo, D. Fan, Ultrahigh-speed rotating nanoelectromechanical system devices assembled from nanoscale building blocks. Nat. Commun. **5**, 3632 (2014)

Printed in the United States
By Bookmasters